Disaster Risk Reduction

Methods, Approaches and Practices

For further volumes:
http://www.springer.com/series/11575

About the Series

Scope of the Series

Disaster risk reduction is a process, which leads to the safety of community and nations. After the 2005 World Conference on Disaster Reduction, held in Kobe, Japan, the Hyogo Framework for Action (HFA) was adopted as a framework of risk reduction. The academic research and higher education in disaster risk reduction has made/is making gradual shift from pure basic research to applied, implementation oriented research. More emphasis is given on the multi-stakeholder collaboration and multi-disciplinary research. Emerging university networks in Asia, Europe, Africa and Americas have urged for the process-oriented research in disaster risk reduction field. Keeping this in mind, this new series will promote the outputs of action research on disaster risk reduction, which will be useful for a wider range of stakeholders including academicians, professionals, practitioners, and students and researchers in the related field. The series will focus on some of emerging needs in the risk reduction field, starting from climate change adaptation, urban ecosystem, coastal risk reduction, education for sustainable development, community based practices, risk communication, human security etc. Through academic review, this series will encourage young researchers and practitioners to analyze field practices, and link it to theory and policies with logic, data and evidences. Thus, the series emphasizes evidence based risk reduction methods, approaches and practices.

Series Editor

Editorial Advisory Group

Rajib Shaw
Editor

Community Practices for Disaster Risk Reduction in Japan

Springer

Editor
Rajib Shaw
International Environment and Disaster Management
Graduate School of Global Environmental Studies
Kyoto University
Kyoto, Japan

ISSN 2196-4106 ISSN 2196-4114 (electronic)
ISBN 978-4-431-54245-2 ISBN 978-4-431-54246-9 (eBook)
DOI 10.1007/978-4-431-54246-9
Springer Tokyo Heidelberg New York Dordrecht London

Library of Congress Control Number: 2013956819

Printed on acid-free paper

Springer is part of Springer Science+Business Media (www.springer.com)

Preface

Community-based risk reduction is widely considered to be important, and several disasters have shown the effectiveness of community responses. Community-based disaster-related activities are termed differently over time. Over more than 100 years ago, before the existence of most states, people or communities were taking care of themselves through collective actions during disasters. After the formation of the state, government-based disaster risk-reduction programs started but failed to fully serve the needs of the people and communities. Over the last 20–30 years, we have been talking again about the need for community-based disaster risk reduction.

Japan is traditionally a disaster-prone country. In older days, Japan used to have a strong community-based approach to cope with disasters. In several places, these practices are still carried out. However, over time, Japan has changed to a strong government-based approach in disaster risk reduction. Several aspects of the infrastructure were developed to protect people and communities from disasters. Innovative research has been done on earthquake risk reduction, especially on high-rise buildings and other structures. However, in 1995, in the Kobe earthquake, more than 6,400 people lost their lives, and it was observed that people and communities were the first responders in the post-disaster scenario. This event changed the scenario of disaster risk reduction in Japan, and innovations in community-based approaches are increasingly being recognized and encouraged.

This book is an attempt to analyze some of the innovative practices of community-based risk reduction in Japan. The book is divided into three parts. Part I describes the evolutionary trend of community practices in Japan. Part II sets forth some of the key issues and sectors where community-based risk reduction has been prominent. Part III provides some case studies from urban and rural areas of Japan, citing various innovative approaches.

This book is intended for researchers and students in higher education in disaster risk reduction. It is hoped that they will find the book useful.

Kyoto, Japan Rajib Shaw

Contents

Part III Case Examples

Contributors

Miwa Abe Kansai University, Osaka, Japan

Kumiko Fujita UNESCO, Jakarta, Indonesia

Jun'ichi Hibino FMYY, Kobe, Japan

Nozomi Hishida Graduate School of Global Environmental Studies, Kyoto University, Kyoto, Japan

Kenji Isayama Department of Medical Science and Technology, Hiroshima International University, Hiroshima, Japan

Shimpei Iwasaki Department of Environmental Sciences, Fukuoka Women's University, Fukuoka, Japan

Junko Mimaki Japan International Cooperation Agency [JICA], Tokyo, Japan

Shohei Matsuura Graduate School of Global Environmental Studies, Kyoto University, Kyoto, Japan

Chiho Ochiai Graduate School of Global Environmental Studies, Kyoto University, Kyoto, Japan

Rajib Shaw International Environment and Disaster Management, Graduate School of Global Environmental Studies, Kyoto University, Kyoto, Japan

Yasutaka Ueda SEEDS Asia, Kobe, Japan

Part I
Evolution of Community Based Risk Reduction in Japan

Chapter 1
Disaster Risk Reduction and Community Approaches

Rajib Shaw

Abstract Community based practices are proven to be effective and useful in disaster risk reduction. This is exemplified not only for Japan, but for many other countries across the globe. Apart from the individual and family, the neighbours in the community are considered as one of the first responders. The community-based disaster risk reduction (CBDRR) approach has been taken by NGOs as a common approach to build resilient communities in their DRR efforts. The approach has been initially implemented in the developing world by NGOs followed by international organizations like the International Federations of Red Cross and Red Crescent. The approach is now increasingly promoted among local governments in order to strengthen the links between the official disaster management system and community-based organizations. There are many case studies of DRR projects with community-based approaches by NGOs and local governments, and there are many variations as well. To ensure sustainability in the community based approaches, it is important to link these to local government programs and policies, especially to different sectors like health, sanitation, education, housing, livelihoods etc.

Keywords Community • Governance integration • Local governments • Risk reduction • Sustainability

1.1 Introduction

It is argued in substantive literatures that disaster risk reduction (DRR) is important and essential in local level. Almost 25 years back, Maskrey (1989) made strong arguments for community based approach in disaster management. After that,

---

R. Shaw (✉)
Graduate School of Global Environmental Studies, Kyoto University,
Sakyo-ku, Yoshida-Honmachi, 606-8501 Kyoto, Japan
e-mail: shaw.rajib.5u@kyoto-u.ac.jp

R. Shaw (ed.), *Community Practices for Disaster Risk Reduction in Japan*,
Disaster Risk Reduction, DOI 10.1007/978-4-431-54246-9_1, © Springer Japan 2014

different literatures and cases of disasters documented, argued and advocated for risk management at the local level (Victoria 2003, 2009; Shaw and Okazaki 2003; Delica-Willson 2005; Kafle and Murshed 2006; Shaw 2012a). However, the definition of local varies from authors, context and countries. Some people argue that anything below national can be termed as local, however, in some cases, it is the local level governments where the focus should be, and in other cases, it is the sub-city or village or community level, where the emphasis of the risk management should be. Policies and practice on DRM, which is the key theme of this chapter needs some attention on this regard. While policies can be made in the government level (at administrative level: either province or city), the practices need to be taken at the community and household levels. Thus, in defining the local DRM, there needs to be a clear link between the local governments and local communities, irrespective of the country and context in the Asia Pacific region.

Community based disaster related activities are termed differently over time. Over more than 100 years ago, before the existence of most of the states, people or communities were taking care of themselves through collective actions during the disasters. After the formation of state, government based disaster risk reduction program started, which failed to serve the needs of the people and communities. Over last 20–30 years, we are now again talking on the need of CBDRR. Thus, community based approach is not new. Rather, we are going back to the old and traditional approaches of risk reduction. Community based disaster management (CBDM) has been a popular term in later 1980s and 1990s, which gradually evolved to CBDRM (community based disaster risk management), and then to CBDRR (community based disaster risk reduction). CBDRM and CBDRR are often used with similar meaning, with enhanced focus on "risk," however there still exists a thin line of distinction. While CBDRR focuses more on pre-disaster activities for risk reduction by the communities, CBDRM focuses a broader perspective of risk reduction related activities by communities, both during, before and after the disaster (Shaw 2012b).

The community-based disaster risk reduction (CBDRR) approach has been taken by NGOs as a common approach to build resilient communities in their DRR efforts (Izumi and Shaw 2012). The approach has been initially implemented in the developing world by NGOs followed by international organizations like the International Federations of Red Cross and Red Crescent (Benson et al. 2001; Maceda et al. 2009). The approach is now increasingly promoted among local governments in order to strengthen the links between the official disaster management system and community-based organizations (Kafle and Murshed 2006). There are many case studies of DRR projects with community-based approaches by NGOs and local governments, and there are many variations as well (Heijmans 2009).

According to Heijmans (2009), towards the end of the 1990s, the approach became an alternative to top-down approaches in disaster management. With this approach, it succeeded to raise people's awareness of disaster risks by using intimate local knowledge, and recognizes pre-existing local capacities and institutions. Therefore, it became possible to improve the position of impoverished, vulnerable, disaster-affected people by addressing the root cause of their vulnerability, and by recognizing their fundamental right to participate in decisions that impact on their lives. At the same time, the capacity development of the community is also critical in this approach to assess the risk, identify risk reduction measures and plan and implement the measures (Delica-Willson 2005).

As social capital (e.g. strong ties, networks and trust) among people is regarded as a key feature to support the well-functioning and resilience of a community in times of a disaster (Murphy 2007; Shaw and Goda 2004), the establishment of community-based organizations during pre-disaster times may have the potential to implement CBDRR (Patterson et al. 2010). However, the efficacy of community-based organizations to implement disaster risk reduction activities may depend on two factors: firstly, the participation level of people within a community; and secondly, the relationship between such a community and the government it belongs to. For example, in an autocratic political system the level of people's participation in decision-making processes is fully dependent on the terms defined by the ruling government, and thus people are likely being manipulated and may not participate (Matsuoka et al. 2012).

The opposite would be that citizen gain full control in decision-making processes and do not rely anymore on an authority or government, which may prescribe certain rules. Since this latter scenario demands a high control within communities to function effectively, in order not to end up in anarchy, governments throughout the world take the lead in guiding communities. However, what is the most favorable relationship between these two actors in relation to implementing DRR? From the perspective where communities are regarded as valuable sources for sharing indigenous knowledge, a certain level of power given to communities may have positive impacts in the overall decision-making process (Matsuoka et al. 2012).

1.2 Primary Issues

Following part describes a few primary issues, which shows the need of the local DRM policy. These issues can be natural context (location and nature of hazards), social context (like diversity of community), institutional context (related to the national and international priorities and sustainability issues) etc.

1.2.1 Changing Nature of Disasters

The nature of disasters, especially the hydro-meteorological disasters is changing, and becoming more as a local phenomena (especially in terms of rainfall pattern). In last several years, catastrophic rainfall is causing serious damages in both urban and rural areas. In Japan, in 2011, there was catastrophic rainfall in Mie and Wakayama prefecture, and one typhoon event brought 1,400 MM of rainfall, almost 80 % of the average annual rainfall of the area. This event kept the local residents unprepared. The communities were cut off from the external assistances, and were left with themselves to respond. Similar incidence was observed in some urban areas, including in Nagoya in 2000, in Mumbai in 2005, in Jakarta in 2007 etc. These are some of the instances of shock events (like typhoon, catastrophic rainfall etc.). The same phenomenon is also observed for the stress events (like drought). In their book on Asian Monsoon drought, Shaw and Nguyen (2011) pointed out that the concept of drought in the Asian monsoon region is very much local. This is more a water management issue, rather than perennial drought in the Africa region. Drought is occurring in parts of Bangladesh, Vietnam, Cambodia; the countries, which are known for its regular annual flooding. Thus, the local nature of disaster is becoming quite prominent due to changing pattern of the rainfall, which is often related to the climatic changes. This urges the increasing need of local capacities (both at the government, non-government and community levels) to cope with theses disasters.

1.2.2 Diversity of Communities

It is a well-accepted fact that the community varies from places to places and its perception and ways to respond to disaster also varies. Therefore, it is important to decentralize the policy and to customize it based on the local needs and priorities. Even within the same city, the analysis of Nakagawa and Shaw (2004) reveals that the social capital changes over place to place. In Kobe, in the district of Mano, the social capital was properly utilized in the recovery process due to a strong community leadership, which helped in the faster decision-making compared to other areas of the same city. Similar observation is also found in case of Gujarat earthquake recovery (Nakagawa and Shaw 2004), which pointed out relatively faster and better quality recovery with a specific community group as compared with others. For a large country in Asia like China, India, Indonesia, different provinces have different cultural, socio-economic and ethnic context. Thus, the risk reduction activities also need to be customized based on the local context. A comparative analysis (Ochiai and Shaw 2009) of Aceh Tsunami (2004) and Yogyakarta earthquake (2005) shows that Yogyakarta had the unique community participation culture (locally called Gotang Royang or mutual help), which was very effective in the recovery process. Thus, the role of housing facilitators was different in case of Yogyakarata (housing facilitators were used in technical problem solving) than that of Aceh (where housing facilitator were used for social problem solving). Thus, the community character defines the level of community involvement and participation.

1.2.3 National Local Linkages

Needless to say, community or local DRM is not independent of national level activities. In fact, there has been increasing awareness both in the country as well as international institutions, that the local DRM should be strengthened with the institutional support from the national governments. The recent Disaster Management Acts of different countries (India, Pakistan, Indonesia, Philippines and Bangladesh) focuses on the issues of legal framework for local DRM. It calls for capacity building of local authorities, enhancing funding for local institutionalization, and create partnerships among different stakeholders at local level.

1.2.4 Evidences from Past Disasters

There are increasing evidences from the recent disasters that the well ware and well prepared local governments and local communities can minimize the impacts of disasters. Japan is a classic example, which was had a tradition central government dominated disaster management system. However, in case of Kobe earthquake of 1995, it was the local communities and neighbors, which helped in 98 % of the survivors, and rest 2 % were rescued by the formal rescue operation. After the Kobe earthquake, Japan's focus was not only to enhance community preparedness, but to make networks of local governments in the form of alliance, to help each other in time of disaster. An outcome of the alliance was to help the city of Toyooka after the 2004 typhoon disaster to overcome the disaster debris and waste by the neighboring cities and prefectures (UNEP 2005). After the 2011 East Japan Earthquake and Tsunami, the Union of Kansai local Governments (Kansai Union) provided support to dispatch of experts and personnel, provision of emergency relief goods, and temporary accommodation of affected populations. Kansai Union is composed of seven prefectures (Osaka-fu, Kyoto-fu, Hyogo-ken, Wakayama-ken, Shiga-ken, Tokushima-ken and Tottori-ken) in the Kansai region.

Similar observations were also found in many disasters in the developing countries. The famous Cyclone Preparedness Program (CPP) of Bangladesh focuses on the capacity building of local residents as the volunteers, keeping in mind that the local communities are always the first responders. In 2005, in Kashmir earthquake, it was the local communities who made the majority of search and rescue, since most of the roads were damaged due to earthquake induced landslides, and thereby affecting the access of the formal search and rescue team.

1.2.5 Increasing Global Awareness on Local Needs

Over past two decades, there has been increasing global and regional awareness on the effectiveness of local needs and priorities. Most of the global and regional framework, including the HFA (Hyogo Framework for Action) urges for local capacity building and policy-making. The Report for the HFA Mid-term review (UNISDR 2011a) admitted that there was still an insufficient level of implementation of the HFA at the local level. In addition, GAR (Global Assessment Report) 2011 mentions that the central role of local governance in DRR and DRM is acknowledged by most countries, and also added that a failure to strengthen local governments and make progress in community participation means that the gap between rhetoric and reality is widening (UNISDR 2011b). This gap is being targeted through international initiatives such as the ISDR World Campaign for DRR "Making Cities Resilient 2010–2015" (UNISDR 2010), which promotes local governments from around the world to take action in implementing DRR activities.

1.2.6 Sustainability and Up-Scaling Issues

Community involvement often faces the problem of sustainability over a longer period of time (Shaw and Okazaki 2003). Government, non-government and international organizations implement various programs before and after the disasters. Many of them are very successful during the project period, however, some of them gradually diminish as the years passed. There are many reasons for gradual decrease of people's involvement in a project. The most common elements are partnership, participation, empowerment and ownership of the local communities. Unless the disaster management efforts are sustainable at individual and community level, it is difficult to reduce the losses and tragedy. While people should own the problems and, consequences and challenges of any mitigation and/or preparedness initiative, it is necessary to see people's involvement in a broader perspective, which is related to policy and strategy. Continuation of community activities over a longer period of time needs a policy environment at local level, as well as local institutions to continue the activities. Thus, even though the initiatives are started with the NGO interventions (as most of the community based activities are observed in Asia), it is important to link them to the local government activities, and incorporate them into policies to ensure its sustainability and replication of innovative efforts to other parts of the disaster prone areas.

1.3 Cross Cutting Issues

With the incorporation of disaster risk reduction in the community level, this section reviews some of the critical issues and ways to implementing risk reduction in the local context. Relation to some of the other cross-cutting sectors is explained with examples.

1.3.1 Link to Habitat

Housing sector is one of the key issues at local and community level, which needs attention in terms of vulnerable housing. In the post disaster context, shelter construction has been a major area where specific aspects of the local DRM can be implemented. There are several good and bad practices exist on this issue after major disasters (Davis 1978; Jha 2010; IRP 2008). While minimum standards of shelter reconstruction have been laid out in the Sphere Standards, such efforts are still grossly inadequate in delivering housing that people want to live in. Numerous agencies work in the affected areas for shelter provision in the aftermath of major disasters. Different approaches are adopted in different areas, with varying materials, sizes and processes. There are, however, certain basic guiding principles for immediate shelter after disaster that must be followed, whatever the approach taken (Sharma 2009).

It is important to identify several shelter related issues like: who is deciding on shelter approaches, where is the expertise, what is the popular wisdom on shelter, what are the dilemmas and conflicts. Use this information to steer shelter recovery programs towards local appropriateness. Location of shelter or recovery housing is important, as well as building material and layout. Rural and urban setting needs different policies and different implementation mechanisms.

Social issues in the shelter construction are also important. In Japanese context of post disaster response, temporary housing plays an important role. Immediately after the disaster, people take shelter in the evacuation center in public facilities like schools or public halls. Depending on the nature of the disaster, it may vary for 3–4 days (for typhoon or rainfall related disasters) to 3–6 months (for earthquake or tsunami). As per the government regulation, the local governments (city and/or prefecture) need to provide temporary housing for the affected people. The duration of the temporary housing is up to 3 years, however, in the Great Hanshin Awaji Earthquake (commonly known as Kobe earthquake), people lived in the temporary housing till 5 years. There are several issues reported in temporary housing in Kobe due to its nature of alignment (closely placed grid pattern, which affect privacy of people), heat and sound insulation problems (which has both physical and mental effects), new neighbors (which prohibits people to come out of their rooms) etc. (UNCRD 2003; Shaw and Goda 2004). Volunteers played a very important role in socialization process in the temporary housing by organizing different events among different age groups. Thus, in post Kobe disasters, special care has been taken to reduce the risk of isolation in the temporary housing through volunteer mobilization.

The World Bank's housing recovery project had a strong component of community involvement, which was embedded in the Kechamatan (village) Development Program (Shaw 2008). The study by Ochiai and Shaw (2009) revealed that the in a devastated community, where several people left their hometown, community building during housing reconstruction was important. The role of housing facilitators (who are the external people with professional expertise in engineering, architecture and social science) has been found to be effective in the community development process.

1.3.2 Link to Health Related Issues

Health and DRM needs to be focused on different aspects in the local DRM policy. There are several complex issues under the broad umbrella of health related issues. It varies is post disaster and normal time. While, it is important to focus on more sanitation and water borne disease issues in most of the developing countries (some of them are considered as a result of climate change), the focus on medical team in the post disaster rescue and relief (both physical and mental health) is also an important health related aspect.

Medical assistance teams during the post-disaster acute and sub-acute phases provide medical care mainly at evacuation centers, and mental healthcare teams are dispatched to the affected areas. Many victims are rescued and supported by them. However, the survival rate is very low 72 h after a disaster event. Survivors may feel changes in the mind and body caused by various disaster stresses, which become magnified as time passes. Disaster causes mental stress and psychological stress occurs as a normal reaction to an abnormal situation. The stress response appears in the body, and thoughts, emotions, and actions change over time. Stress reactions in victims of a disaster weaken gradually over time during disaster reconstruction. Adverse effects often appear because of the traumatic experience caused by the disaster. These reactions include hyper arousal, re-experiencing the event, and avoidance. Healing spontaneously within 1 month from such a state is called acute stress disorder (ASD); if the symptoms persist for >1 month, it is called post-traumatic stress disorder (PTSD). In contrast, stress responses during a disaster appear not only in victims but also in the relief referred workers who are referred to as "hidden victims" (Isayama and Shaw 2013).

Human beings are also exposed to climate change through changing weather patterns and indirectly though changes in water, air, food quality and quantity, ecosystems, land use, agriculture, livelihoods and infrastructure. These direct and indirect exposures can cause death, disability and suffering (Kien and Shaw 2009). Lower health state of a nation, a group or an individual would increase vulnerability and reduce the capacity of the individual or group to adapt to climate change. Recently, the Fourth Assessment Report of the IPCC had found the newest evidences and suggested with very high confidence that climate change has already significant impacts on human health and contributed to the global burden of disease and premature deaths. Climate change has already altered the distribution of some infectious disease vectors; altered the seasonal distribution of some allergenic pollen species and increased the heat wave-related deaths. Extreme climate events include heat waves, heavy rains and flooding, droughts, typhoons and storms and others. Over time, regional and local populations have been adapting to such local prevailing climate events, both passively (autonomously) and actively (planned adaptation) due to physiological, behavioural, cultural and technological responses and measures. However, extreme events often stress populations beyond those adaptation limits. Understanding the health risks from these events is important because the future frequency and intensity of extreme events is expected to increase.

Another contrasting issue is health care in the aging society. According to the 2010 Population Census, which serves as the base year of these projections, the total population of Japan in that year was 128.06 million (total population including non-Japanese residents). Based on the results of the medium-fertility projection, Japan is expected to enter a long period of population decline. The population is expected to decrease to around 116.62 million by 2030, fall below 100–99.13 million in 2048, and drop to 86.74 million by 2060. Japan obtained universal public pension and health insurance coverage in 1961 (National Institute of Population and Social Security Research 2010). Long-term care for older adults has been a priority for the past two decades. However, a chronic shortage of rural physicians continues, and elderly adults have not benefited enough from modern western medicine. Public health nurses must support the health problems in rural regions. By reviewing some of the recent disasters in Japan, Isayama and Shaw (2013) propose a healthcare networking system, which takes care of their daily healthcare needs through community and neighbourhood networking, and this, helps in responding to disaster effectively. This issue needs to be strongly incorporated in the local DRM policy and practice through incorporating health care training for the resident association (locally called jichikai in Japanese) and self-help disaster prevention committee (jishubo in Japanese).

1.3.3 Link to Livelihoods

It is evident that the local economy gets affected when a natural hazard or a disaster strikes. The disasters are detrimental not only to the lives, but also to the source of livelihoods and productive assets. Disaster impacts in developing regions are potentially grave as they affect agriculture, sanitation, and lack of water-food supplies, and the consequent loss of living space and livelihood could lead to the migration of people towards "safer" places. Direct economic loss of disaster is obvious, and much talked. However, in this section, emphasis is given on the small-scale livelihoods, where people find it hard to recover from the mental abasement and rebuild their assets to earn. Therefore, the damage is maximum for self-sustaining small scale industries relying on indigenous infrastructure (Srivastava and Shaw 2012). Means of living such as farming, fishing, small trade, micro and small enterprises in the formal and informal sector, all get paralyzed. Even the enterprises running on loans find it hard to repay it with a halt to income generation post disaster. This could lead to virtual collapse of the economic and social environment. The impact of a disaster on the local economy can be comprehended by the case of Bhuj earthquake 2001 in the Kutch region of the Gujarat state, India (PNY 2003; Shaw et al. 2003). The damage to plants, factories and machinery caused a shutdown to approximately 10,000 small and medium scale industrial units. For example, the major victims were diesel engine manufacturing and machine and tools industry in Rajkot, ceramic units in Morbi and Surendranagar, art and small tools industry in Kutch and thousands of salt pans. Moreover, more than 50,000 crafts persons who lived and worked in Bhuj, Anjar, Rapar, and Hodka, lost their livelihood due to the damage (Vatsa 2001).

Apart from the shock events like earthquake, tsunami, cyclone or typhoon, there are stresses occurring in different areas, which also affect the migration of people from the rural to urban areas in search of new livelihoods. Urban rural linkage has a very strong rationale behind considering rural–urban interactions in understanding employment and disaster together is to reason out how changing rural–urban interactions affect the livelihoods of low income and vulnerable groups in urban and rural settlements. The extractive relationship of urban area towards rural areas needs to be changed into a network relationship where urban and rural areas can together sustain the stress caused by disasters. Also, the role of community needs to be strengthened to build their occupational resilience, including adopting appropriate technologies while making the most of traditional knowledge, and diversifying their livelihoods to cope with current and future climate stress and develop "Disaster proof activities" (UNFCCC 2009).

1.3.4 Link to Education System

Local DRM needs to be strongly linked to formal and non-formal education system. Realizing the crucial role of DRR (disaster risk reduction) education, there is a need to focus on implementation of DRR education in terms of safer schools, enhancement of resilience capacity, and reduction of losses from disasters. As the inherent attribute of education lies in the inter-relation from management level of policy-makers and school managers to practice level of students and communities, it is important to develop a comprehensive approach to promoting DRR education at both policy and school levels. Besides, when considering DRR education, integration of DRR into education curricula only is not enough to bring about meaningful risk reduction. It should also include related issues such as structural and non-structural safety, legislative basis, management mechanism, qualified human resources, sufficient funding, strong collaboration, proper warning system and risk assessment, among others. Thi et al. (2012), in an innovative approach to assess the resilience of education sector have identified that the assessment should include physical conditions, human resources, institutional issues, external relationships and natural conditions. In policy approach, Gwee et al. (2011) pointed out that 16 tasks of HFA (Hyogo Framework for Action) can be regarded as important for the education sector, and is considered as E-HFA (HFA in educations sector). Both these approaches need to be linked to the DRM practices in the local level, especially in close cooperation with education board of the city and or local governments.

1.3.5 Link to Environmental and Natural Resource Management

With increasing dependence of local communities (especially in the rural and semi-urban areas), the local DRM needs to be linked to sustainable use of natural resources. Environmental conditions may exacerbate the impact of a disaster, and vice versa, disasters have an impact on the environment. There are many adverse impacts of the environment degradation on human vulnerability and disaster, among which the impacts of deforestation, forest management practices, agriculture systems, etc. exacerbate the negative environmental impacts of a storm or typhoon, leading to landslides, flooding, silting and ground/surface water contamination (Tran et al. 2009). Therefore, a sound local DRM should be linked to the local environment and natural resource management policy (UNEP 2005).

First, the rapid loss of forestland around the world is changing the rainfall patterns. Deforestation is not only linked with decreased rates of evapo-transpiration, but also a reduction in moisture flow and water retaining capacity. These changes, have been linked with the onset of large-scale deforestation, and have led either to drought and desertification or soil erosion, debris flow and floods. In fact, river and lake floods are aggravated by deforestation, which, in turn, causes erosion and clogs rivers. Water quality and quantity in the downstream areas are heavily affected due to the damage of the vegetation in the catchment areas, which reduce agricultural productivity in the downstream areas (Le and Rambo 1996).

Studies show that river catchment areas that are largely deforested or wetlands that have been drained create very different hydrological regimes (Gujja 2001). When this factor is added to the climate change, it appears to be affecting the timing and pattern of the rainfall. Eventually, the rainfall may become more erratic. In this regard, the timing and volume of expected flooding are becoming more unpredictable. Impacts of deforestation are being felt more severely in the highlands, but they will eventually affect the lowlands: as rivers silt up, this affects croplands in the lowlands that depend on these waters. Flooding of cities will become more pronounced. According to Wang (2004), it has seen many "natural hazards" happening frequently due to environmental degradation recently.

Second, the non-sustainable over-use of resources causes pollution and ultimately leads to environment degradation. In particular, there is an increasing likelihood of human induced climate change which according to the latest projection of the Intergovernmental Panel on Climate Change, will result in more water related disasters especially for countries in tropical and sub-tropical latitudes (WMO and GWP 2006). These changes in temperature and related local rainfall variations affect the environment through accelerated desertification, land degradation, the availability of water resources as well as reducing the overall agricultural output. In addition, climate change is expected to affect sea levels and cause climate extremes. All these factors have a compound effect on the occurrence and impact of disasters. Uy et al. (2012) from an illustrative example from Infanta, Philippines has pointed out the need of eco-system based adaptation measures to be incorporated in to the local (barangay or village) and municipal level DRM plans policies.

Third, many mangrove swamps and coastal forests are under severe threat from various river basin development activities and flood management projects. Flood prevention projects distort the flow variability and can cause severe deteriorations of the coastal ecosystems. The degradation of mangrove forests significantly reduces their valuable functions such as shoreline stabilization and storm protection. The coastal forests become too degraded to absorb the energy of coastal storms, thereby increasing the flood and storm risks in the coastal zone (Tran and Shaw 2012).

1.4 Policy Issues Related to Community Practices in DRR

Based on the above observations, a few generic policy recommendations for implementing DRR policies and practices are as follow:

Local DRM needs to be linked to development issues like health, education, and sanitation: It is always a challenge to continue the DRM activities at the local level for a longer period of time. This sustainability issue is often discussed and analyzed. There are several models of sustainability of local DRM, linking it to the local communities, or providing ownership to the communities, or depending on the local leadership etc. After the disaster, community based DRM becomes popular, and after a few years, the initiative gradually diminishes. Thus, to sustain the local DRM over a longer period of time, it is required to link it to the daily needs, like welfare, health, sanitation, environment etc. The nature of the community or local context defines the specific entry point for sustainability of DRM activities. For example, in case of urban areas in developing countries, environmental entry points like solid waste management, sanitation, water issues are possible areas where the local DRM can be linked. For the rural communities, possibly the health and welfare activities are the appropriate entry points for local DRM.

Resource commitments of the local governments are required: One of the key points of local DRM is the policy provision for utilizing appropriate budget and funding. Preferably, a certain percentage of local development budget needs to be allocated for local DRM. If difficult, specific activities like regular capacity building programs or awareness-raising programs need to be budgeted properly in the local level.

Regular updating and testing of local DRM initiatives is essential: Community is dynamic, and there are changes in the communities. Thus, the local DRM needs to be dynamic as well. The approaches may be same, but needs to be customized based on the local changes and context. The technology and governance system needs to be adjusted accordingly. Also, the traditional knowledge needs to be transferred over generation and context.

Strengthen linkage with local knowledge institutions, like university is useful: The link with the local institutions is important for sustaining the knowledge base of local DRM. The local university can be a resource organization to provide technical expertise to the local governments and communities, and thereby making a sustainable knowledge management system.

1.5 About the Book

With the above context of general issues and approaches of community based disaster risk reduction, this book focuses specifically on Japanese examples. Japan has incorporated traditionally a government-oriented approach over the last century. To be more precisely, it was the community-based approach, which was turned to government-based approach to ensure safety and security of people and assets. It was until 1995, that the government-based approach dominated, along with engineering focused risk reduction measures. However, the Kobe earthquake pointed out that it was the local communities and their neighbours who did maximum rescue operation. Also, the major damages were concentrated in vulnerable groups of aged population. This disaster was the turning point of community-based approaches of Japan, starting from the evacuation centre to temporary housing and then to permanent recovery housing.

Over last 18 years, different community based innovations are implemented in different parts of Japan, to cope with different types of disasters. Many policy changes and regulations also encourage community based risk reduction approaches. This book analyses them in to three parts with 13 chapters. The first part talks about evolution of the community based approaches with three chapters. The second part focuses on issues and challenges with four chapters. The third and final part presents six case examples from different types of disasters.

This chapter covers evolution of CBDRR approaches, its issues, challenges, linkages to different cross cutting issues, and a few policy perspectives to implement CBDRR in a sustainable way. Chapter 2 focuses on the changes happened in community based practices in Japan after the Great Hanshin Awaji Earthquake of 1995, which is also termed as the Year Zero of volunteerism in Japan. The concept of civil society has changed from anti-government (watch-dog type) to cooperative actions with the government. The event is not only important to the history of Japan, but also to the world in terms of community based practices. The strong government oriented search and rescue system could rescue only a handful number of people, where most of them were rescued by family and neighbours. Chapter 3 focuses on community involvement in different types of disasters related to sediment disaster, typhoons and flood. The chapter starts with 1999 Hiroshima disaster and explains the process of sediment disaster law, and the importance of community involvement. Year 2004 was a year of typhoon in Japan, when 13 typhoons hit mainland Japan. In typhoon 21 and 23 of 2004, major devastation was observed in different parts of Japan. It showed that while the traditional typhoon preparedness (like early warning, evacuation, shelter) worked somehow well, the key issue is in the forest management in the mountain areas, which caused significant landslides and mass movements. Thus, the upstream downstream linkages at the community level became important and gradually being recognized widely. This chapter describes a unique education program, and roles of volunteers, which links the school, community and the local governments.

Chapter 4 makes an historic analysis of role of social capital in disaster risk reduction in Japan. Generally, Social Capital refers to networking, trust, and norms among people, which can be formed in the society. It is supposed that being rich in social capital, in other words, having strong faith among citizens (or members which configure a community) and having strong ties inside communities positive effects, such as enhancing political commitment, improving performance of government and businesses to be more effective, improving educational performance, public security and stimulating the local economy. Although, social capital can give negative impacts to recovery processes sometimes, generally the strong state in social capital can contribute to the recovery of communities affected by disasters. That is because social capital facilitates various activities and enhances performance of social actors, which are essential for disaster recovery.

Mega disasters, such as the East Japan Earthquake and Tsunami (EJET), have devastating effects on schools and communities. Not only do they interrupt educational activities, but also weakens community ties due to displacement of affected people that exacerbates the chronic low birth—aging problem faced by many rural communities in Japan. Chapter 5 rationalizes the concept of School Centered Community Building that is proposed by the Ministry of Education, Culture, Sports, Science and Technology (MEXT) of Japan as a possibility to restore the community ties weakened by EJET though schools, by first looking into new ways to utilize the schools by making them multi-functional facility aimed to strengthen school—community linkage. It then examines the approaches of community participation in school management and educational activities to realize school-centered community building. Finally, it analyzes the benefits and challenges of this concept to come up with possible ways to advance the school recovery process.

Physical and mental healthcare are needed during the reconstruction/rehabilitation phase after the disaster as well as in normal daily life. However, "earthquake-related deaths" were reported due to overwork and deterioration from chronic diseases after a disaster, and the insufficiency of the healthcare network as one of the factors is pointed out. It is necessary to strengthen the community network through a daily healthcare network with cooperation among social welfare, medical, and local government agencies at multi levels to maintain good healthcare for local residents after disasters, and to promote health. Chapter 6 elaborates this concept of healthcare preparedness by the local communities, which enhances disaster risk reduction.

There was one community radio station established by the residents themselves immediately after the Great Hanshin Awaji Earthquake in 1995 to provide disaster information to the affected people and to heal their psychology trauma. It was discovered that community radios could play an important role more than mass media radio and TV stations. This led to community radio stations springing up here and there in Japan. As a result of these radio stations being established, bonds within communities became stronger which in turn led to an increase in community DRR capabilities in every natural disaster. Chapter 7 focuses on the role of community radio for community-based disaster management.

High aging ratio and depopulation are observed in almost all Japanese rural area. Chapter 8 focuses on Tosashimizu city, Kochi, which is located in rural area. Owing to its physical and societal vulnerability against natural disasters, the local government has promoted activities for community based disaster management (CBDM). Efforts by residents show that CBDM contributes to not only reduce disaster risk but also revitalize even aging society.

The attention given to community-based disaster management (CBDM) has grown lately due to the increase in the number of natural disasters. People have been fighting disasters such as earthquakes, floods, and fire to protect their lives, houses, and villages. Since most of the houses in Japan are made of wood, preventive measures against fire are prioritized. Historical and cultural houses and buildings often attract people's attention. However, more importantly, that cultural heritage can be understood to be the result of efforts by local residents who have been maintaining structures and handing down traditions and beliefs to protect these heritage sites over the generations. Shirakawa village, a UNESCO World Heritage Site, is located in an isolated, heavy-snow environment. It has a long history of protecting its traditional houses, which are made with thatched roofs that are vulnerable to fire. The people of this village have formed associations, such as the Residents' Association (RA) and Fire Volunteer (FV) force, and undertaken several activities to keep social order and prepare for disaster and emergency. The study revealed that people in the village have been sustaining their traditional organizations and activities in spite of modernization and socio-environmental changes. Chapter 9 focuses on the processes and mechanisms that have been established by local organizations and activities to protect the Gassho-zukuri houses, which are prone to fire.

There are two major measures for disaster risk reduction, structural and non-structural measures. Early warning and evacuation are the major non-structural measures, which are based on judgment and action of local people. Their judgment and action are decided by information through various media. In Chap. 10, relation of media and local people in Reihoku area is analyzed. Reihoku area is a landslide disaster-prone area in Kochi prefecture. Possible information system for disaster risk reduction in Reihoku area is discussed based on the result of a questionnaire survey.

Chapter 11 shows the findings obtained from the hearing survey conducted concerning relocation impacts and land use of pre/post relocated site by the relocated people and people living near the relocated area after Chuetsu earthquake, 2004. Research targeted one of the affected community named "Kotaka" which is located in Niigata Prefecture. Kotaka was relocated far from their original land. As a result of relocation, people had conflicts within the community as well as external community. There were also conflicts within own family members because of quick decision-making, conflict with leaders, rumor and misunderstandings of the governmental relocation projects, as well as different opinions between young and old. However, through events such as sports festival, relocation of community temple, traditional dance and so on, it helped bonding within the community. Kotaka's relocation shows importance factors on how to keep community (village) identity after relocation. This way, relocation would not be simply just movement of the people (i.e. extinction) but movement of an entire community.

In Japan, when a large-scale disaster happens, temporary housings are provided by the government based on the law. Accordingly more than 53,000 of temporary housings were constructed in the Great East Japan Earthquake and Tsunami occurred on 11th March 2011. During the Great Hanshin-Awaji Earthquake (1995), communities that had existed before the disaster were fragmented, as a consequence, residents of temporary housing were forced to live with non-acquaintances and in the worst case some isolation deaths happened. Similarly, in the case of this disaster, many residents were assigned their temporary housing by lottery, but without consideration of existing community ties amongst the affected people. Therefore, to build and develop community activities was one of the challenges in temporary housings.

Chapter 12 shows overview of provision of temporary housings in the Great East Japan Earthquake and Tsunami, and case of community activities in temporary housings in Kesennuma city conducted by non-government organization with a local community based organization. In addition, importance of promoting communication among various stakeholders such as residents, generations, and people affected by previous disasters, and of linking local issues found by the communication to aid organizations and local governments is pointed out through the case.

A coastal lagoon which is a body of shallow water separated from the sea by barriers is highly exposed to environmental and climatic factors. The lagoon is characterized by an essential quality of uncertainty for use in resource management, especially in the field of aquaculture. This paper presents a case study of Saroma Lake, Japan where innovative efforts by the fishers have been made on wise use of aquaculture. One of innovative efforts includes recruitment of scientists in their aquaculture cooperative association, which enabled the fishers to adapt to changes in the lagoon environment flexibly and mitigate the impacts of climate change with active participation of relevant stakeholders. The lessons learned from the case study in Chap. 13, provide roles of residential scientists in adaptive risk management toward responsible fisheries, which takes into environmental change account.

Acknowledgements This chapter is based on different earlier work over more than 10 years, which is highly acknowledged.

References

Benson C, Twigg J, Myers M (2001) NGO initiatives in risk reduction: an overview. Disasters 25(3):199–215

Davis I (1978) Shelter after disaster. Oxford Polytechnic Press, Oxford

Delica-Willson Z (2005) Community-based disaster risk management: local level solutions to disaster risks. Trop Coasts 12(1):66–73

Gujja B (2001) Dam accused of role in flooding: research paper: "Dams and Floods". World Wide Fund for Nature. Retrieved 12 December 2012, http://www.reliefweb.int/

Gwee Q, Takeuchi Y, Jet-chau W, Shaw R (2011) Disaster education system in Yunlin County, Taiwan. Asian J Environ Disaster Manag 3(2):189–204

Heijmans A (2009) The social life of community-based disaster risk reduction: origins, politics and framing. Disaster Studies Working Paper 20. Aon Benfield UCL Hazard Research Center, Boulder, USA

IRP (2008) Guidance notes on early recovery. International Recovery Platform, Kobe, 58 pp

Isayama K, Shaw R (2013) Integrated healthcare as the future of disaster recovery potential in Tohoku region. In: Shaw R (ed) Disaster recovery: used or misused development opportunities. Springer, Tokyo, Japan, pp 363–383

Izumi T, Shaw R (2012) Role of stakeholders in disaster risk reduction under Hyogo Framework for Action: An Asian Perspective. Asian J Environ Disaster Manage 4(2):164–182

Jha A (2010) Safer homes, stronger communities. GFDRR, The World Bank, Washington DC, USA, pp 367

Kafle SK, Murshed Z (2006) Community-based disaster risk management for local authorities: participant's workbook. Asian Disaster Preparedness Center (ADPC), Pathumthani

Kien T, Shaw R (2009) Climate change adaptation and human health: inter-linkages of climate, disaster and health issues in Vietnam. In: Shaw R, Krishnamurthy R (eds) Disaster management: global challenges and local solutions. University Press, Hyderabad, pp 444–467

Le C, Rambo T (1996) Red books, green hills: the impact of economic reform on restoration ecology in the Midlands of Northern Vietnam. East-West Center, Honolulu, p 75

Maceda E, Gaillard J, Stasiak EV, Le Masson V, Le Berre I (2009) Experimental use of participatory 3-dimentional models in island community-based disaster risk management. Int J Res Island Cult 3(1):72–84

Maskrey A (1989) Disaster mitigation: a community-based approach. Oxfam, Oxford

Matsuoka Y, Joerin J, Shaw R, Takeuchi Y (2012) Partnership between city government and community-based disaster prevention organizations in Kobe, Japan in community based disaster risk reduction. Emerald, Bingley, pp 151–184

Murphy BL (2007) Locating social capital in resilient community-level emergency management. Nat Hazards 41:297–315

Nakagawa Y, Shaw R (2004) Social capital: a missing link to disaster recovery. Int J Mass Emerg Disaster 22(1):5–34

National Institute of Population and Social Security Research (2010) Population & household projection. Retrieved 30 December 2012, from http://www.ipss.go.jp/site-ad/index_english/population-e.html

Ochiai C, Shaw R (2009) Participatory urban housing reconstruction study in Aceh, Indonesia. In: Shaw R, Srinivas H, Sharma A (eds) Urban risk: an Asian perspective. Emerald, Bingley, pp 233–254

Patterson O, Weil F, Patel K (2010) The role of community in disaster response: conceptual models. Popul Res Policy Rev 29:127–141

PNY (2003) Patanka new life plan. SEEDS, UNCRD, Kobe, 48 pp

Sharma A (2009) Community based disaster recovery. In: Shaw R, Krishnamurthy R (eds) Disaster management: global challenges and local solutions. Hyderabad, India, pp 352–364

Shaw R (2008) Kechamatan Development Program, Indonesia. Case study in building resilient communities. The World Bank, Washington DC, USA, pp 243–267

Shaw R (2012a) Community based disaster risk reduction. Emerald, Bingley, 402 pp

Shaw R (2012b) Overview of community based disaster risk reduction, Community based disaster risk reduction. Emerald, Bingley, pp 3–17

Shaw R, Goda K (2004) From disaster to sustainable civil society: the Kobe experience. Disasters 28(1):16–40

Shaw R, Nguyen H (2011) Droughts in Asian monsoon regions. Emerald, Bingley, 186 pp

Shaw R, Okazaki K (2003) Sustainability in grass-roots initiatives: focus on community based disaster management. UNCRD, Kobe, 99 pp

Shaw R, Gupta M, Sharma A (2003) Community recovery and its sustainability: lessons from Gujarat earthquake of India. Aust J Emerg Manag 18(2):28–34

Srivastava N, Shaw R (2012) Interrelationship of disaster and employment from the perspective of urban rural linkage in India, Environment disaster linkages. Emerald, Bingley, pp 145–164

Thi T, Shaw R, Takeuchi Y (2012) Climate disaster resilience of the education sector in Thua Thien Hue Province, Central Vietnam. Nat Hazards, DOI:10.1007/s11069-012-0178-5

Tran P, Shaw R (2012) Environment disaster linkages: an overview, Environment disaster linkages. Emerald, Bingley, pp 3–13

Tran P, Sonak S, Shaw R (2009) Disaster, environment and development linkages: opportunities for integration in Asia-Pacific region. In: Shaw R, Krishnamurthy R (eds) Disaster management: global challenges and local solutions. University Press, Hyderabad, pp 400–423

UNCRD (2003) From disaster to community development: the Kobe experiences. UNCRD, Kobe, 109 pp

UNEP (2005) Environmental management and disaster preparedness: lessons learnt from the Tokage Typhoon. UNEP, Osaka, 28 pp

UNFCCC (2009) Climate change: impacts, vulnerabilities and adaptation in developing countries. Retrieved 14 December 2012, from http://unfccc.int/resource/docs/publications/impacts.pdf

UNISDR (2010) Information kit for ISDR 2010–2011 world disaster reduction campaign. http://www.unisdr.org/english/campaigns/campaign2010-2011/documents/campaign-kit.pdf

UNISDR (2011a) Hyogo framework for action 2005–2015 building the resilience of nations and communities to disasters, mid-term review 2010–2011. UNISDR, Geneva. http://www.unisdr.org/files/18197_midterm.pdf

UNISDR (2011b) Global assessment report on disaster risk reduction 2011. UNISDR, Geneva. http://www.preventionweb.net/gar

Uy N, Takeuchi Y, Shaw R (2012) An ecosystem based resilience analysis of Infanta, Philippines. Environ Hazards, DOI:10.1080/17477891.2012.688794

Vatsa KS (2001) The Bhuj earthquake. A Reconnaissance Report: identification of priority issues. Retrieved on 29 December 2012, from http://www.drmonline.net/drmlibrary/pdfs/Gujarat_report.pdf

Victoria LP (2003) Community based disaster management in the Philippines: making a difference in people's lives. ADPC, Bangkok

Victoria L (2009) Community capacity and disaster resilience. In: Shaw R, Krishnamurthy R (eds) Disaster management: global challenges and local solutions. University Press, Hyderabad, India, pp 338–351

Wang Y (2004) Environmental degradation and environmental threats in China. Environ Monit Assess 90:161–169

WMO and GWP (2006) Environmental aspects of integrated flood management, Second Draft. World Meteorological Organization and Global Water Partnership-Associated Program on Flood Management, Genera, Switzerland

Chapter 2
Kobe Earthquake: Turning Point of Community-Based Risk Reduction in Japan

Rajib Shaw

Abstract The Great Hanshin Awaji Earthquake [popularly known as Kobe Earthquake] is considered as the turning point of community based approached in Japan. Data shows that communities played the most vital role in immediate rescue operation. Communities also facilitated the temporary housing lives, and short- and long-term recovery. New dimensions of civil society emerged through the recovery program, where a stronger bond of local government and resident association was observed. Kobe city government started new innovation of school-centered community building, linking the social welfare and disaster risk reduction. "Bokomi" [disaster prevention and social welfare community] is considered a unique sustainable approach to engage communities to risk reduction activities and to enhance collaboration with the local governments.

Keywords Bokomi • Kobe earthquake • Local government collaboration • New civil society • School community linkage

2.1 Introduction

The Kobe earthquake with a magnitude of 7.2 on the Richter scale, and with a depth of 16 km hit the city of Kobe and its surrounding areas in Hyogo Prefecture on 17 January 1995 at 5:46 AM. The total number of casualties exceeded 6,400, with numerous injuries and victims of other collateral damages. Buildings and infrastructure were severely damaged, and more than 200,000 people had to find temporary shelter in different parts of the city. Within Kobe city administrative area alone,

R. Shaw (✉)
Graduate School of Global Environmental Studies, Kyoto University,
Sakyo-ku, Yoshida-Honmachi, 606-8501 Kyoto, Japan
e-mail: shaw.rajib.5u@kyoto-u.ac.jp

R. Shaw (ed.), *Community Practices for Disaster Risk Reduction in Japan*,
Disaster Risk Reduction, DOI 10.1007/978-4-431-54246-9_2, © Springer Japan 2014

70,000 buildings completely collapsed, and 55,000 were seriously damaged. Public facilities like city offices, schools and hospitals were also damaged extensively, which rendered the city services paralyzed for several days. Utility services were also interrupted: electricity services were out of order in the entire metropolitan area, 25 % of the telephone services did not work, water and gas services were disrupted throughout the entire city. At several locations, severe fires broke out, and 7,000 buildings were completely burned, resulting in more than 800,000 m^2 of burnt areas. The damage to social and industrial capital stock was estimated at seven trillion JPY within Kobe city. Secondary and tertiary losses in the city and other parts of the prefecture were much higher (Nakagawa and Shaw 2004).

Damage caused by the Kobe earthquake was concentrated on low-income old people living in the inner city, making this the first disaster that an ageing "modernistic megalopolis" in Japan has faced (Miyamoto 1996). Urban area reconstruction faces intricate problems because of its complex nature, and the social, economic and cultural context. After the event, both financial and human resources arrived in the affected areas, however, the challenge was to their sustainability. The rationale for community involvement or community-based activities is now well rehearsed (Twigg 1999). Because community-based activities (and community-based organizations) are deeply rooted in the society and culture of an area, they enable people to express their real needs and priorities, allowing problems to be defined correctly and responsive measures to be designed and implemented. Twigg (1999) also argued that the existence of community-based organizations allows people to respond to emergencies rapidly, efficiently and fairly, and therefore the resources will be used economically. Maskrey (1989) pointed out that "top-down" programs in which communities are not involved tend not to reach those worst affected by disaster, and may even make them more vulnerable. Both in developed and developing countries, at the time of emergency, largely ad hoc responses by loosely knit groups from within communities are observed, before formal organizations are able to mobilize. These groups play important roles in disaster response and reconstruction process (Raphael 1986).

The rehabilitation of Kobe started on 17 March 1995 with the announcement from the Hyogo Governor on "Designation of Land Readjustment and Redevelopment Areas" (*Toshi Keikaku Kettei* in Japanese). The designation was open to public inspection for 2 weeks and residents and concerned persons could object to the plan via written documents. The City of Kobe designated six readjustment and two re-development areas but soon after the announcement many heated arguments arose among the residents from those designated areas in the plan. The designation was in many ways controversial. The decision was made without any consultation with the residents. Although it was open to public inspection, little flexibility was seen on the city/prefecture administration side regarding any changes to the plan. Naturally, the negotiation between residents and the administration became bogged down in some areas and the rehabilitation was delayed (Nakagawa and Shaw 2004).

In the earthquake-affected areas, those designated for land readjustment and redevelopment were termed as "black zones" and other areas were called "white zones" by the stakeholders (Nakagawa 2003). The division depended on the level of commitment and involvement of public agencies toward the rehabilitation. Thus, there were many differences in official support for the rehabilitation in these two zones. In "black zones", property owners needed to make sacrifices for land adjustment or redevelopment to proceed, however, the government provided the physical rehabilitation such as building wider roads and parks; and normally the environment was improved incorporating disaster management aspects. But in "white zones", narrow roads remained narrow and some illegal construction during the confused period made the environment even worse than before. Also, the government provided practically no financial support. In addition, some special preferences were given to "black zones", for instance in the sale of land up to 50,000,000 yen (US$1 = 100 yen in 1995) and the exchange of land were tax-exempted (Kinmokusei International Project 1999).

However, the most important difference was that in every "black zone", "Machizukuri" (Town Development) organizations were formed. A machizukuri organization is an organization consisting of residents, private agencies and others with an interest in the area's restoration. In Kobe, most of the machizukuri organizations were formed based on the existing community organizations such as neighbors' associations. Machizukuri organizations provide very important "opportunities" for community members to discuss future city planning and this was the first step to community participatory rehabilitation. Machizukuri organizations also acted as the interface with city officials and city planning consultants. Consultants and advisors also played a big role in the rehabilitation process. Consultants were dispatched to each machizukuri organization and provided technical and professional knowledge on city planning. In contrast, in the "white zones", the forming of machizukuri organizations was not mandatory since the areas were not designated as the official project locations. In spite of many difficulties, several machizukuri organizations were formed in the "white zones", but many were not officially recognized under the ordinance. There were areas called "gray zones" where these machizukuri organizations had existed before the earthquake for different development projects in the urban areas. However, as in the "white zones", "gray zones" were not entitled to the special preference mentioned earlier for "black zones", this resulted in similar situations as those in the "white zones" (Kinmokusei International Project 1999). Every machizukuri organization faced various difficulties in the reconstruction process, and there were obvious differences in the speed and the degree of people's involvement among the communities. In some areas, negotiation between residents and government was prolonged on issues such as the amount of land that owners in land-readjustment districts should contribute for public improvement, which resulted in an even split of machizukuri organization into several residents groups.

In this chapter, some of the key community based issues and approaches in post Kobe have been discussed. Kobe is often considered as the year zero of volunteer activities in Japan, which is discussed first. The emergence of a new civil society model is discussed from a neighborhood-based approach. Finally, the welfare based model is discussed as a sustainable community involvement model which emerged from Kobe.

2.2 1995: Year Zero of Volunteer Activities in Japan

Japan's non-profit sector has a long history, starting from the Meiji era in 1898 with the enactment of the civil law code, which established regulations for public interest corporations (Matsubara and Todoroki 2003). The main aspect of the public interest corporation system was that government would regulate these organizations. After the establishment of the code, during the last 100 years, the concept of NGOs and civil society has changed, although most of the time it was under the control of the government. Deguchi (1999) pointed out three different conceptions of civil society in Japan. Termed shimin sakai in Japanese, one meaning assigned to it is the modern society that emerged in the wake of the French Revolution as defined by Western liberal political thinkers. Thus, many Japanese associate this term with the French Revolution. The second usage was a by-product of Japanese political ideology after the Second World War. During this era, Japan's leftist intellectuals believed that Japan was moving towards a socialist society in which the leading role will be played by the shimin or citizens. From the end of the Second World War until the 1980s, the term shimin sakai was most commonly used in this sense. The third usage of civil society corresponds to a society in which voluntary organizations or non-profit organizations (NPOs) play an important role. This meaning came into existence in Japan only after the 1995 Kobe earthquake (Shaw and Goda 2004).

After the Second World War, Japan experienced extremely rapid economic growth. Some people argued that this was because of the so-called "Iron Triangle"; the three angles of which are politicians, business leaders and government officials. The triangle worked very well until the 1970s. It is called 'the 1940 System' by some scholars (Noguchi 1995), which turned Japan into a highly centralized, bureaucrat controlled society. However, the system began losing power when the government could no longer offer enough of the requisite social services and proposed their privatization. Major grant-making foundations established in 1970s, such as the Toyota Foundation (1974), Suntory Foundation (1979) and the Nippon Life Insurance Foundation (1979) began to support volunteer activities. The Plaza Accord of 1985 brought a rapid appreciation of the yen, which obliged Japanese business corporations to establish production bases abroad, especially in the US. Through contact with community leaders in America, Japanese business leaders became acquainted with the concept of corporate citizenship (Imada 1999). The period from the late 1980s was one of material abundance in Japan after long economic growth. As a result, Japanese people began to want more spiritual and human satisfaction rather than only material satisfaction. A combination of a gradual awareness among Japanese of community participation and exposure of Japanese business leaders to corporate citizenship, urged the Keidanren (Federation of Economic Organizations) to send a study team to the US in 1988. As a result, the "1 Per Cent Club" was established within Keidanren in April 1990. The 1 Per Cent Club was a voluntary association composed of corporations willing to donate 1 % of their earnings. By September 1997, individual membership in the 1 Per Cent Club reached 1,225, while corporate membership was 281. In 1990, the Osaka Chamber of Commerce and Industry published a report on community foundations. A year later, the Osaka Community Foundation was established by an initiative from the Chamber of Commerce and Industry. In 1993, the Philanthropy Link-Up Forum was established in Osaka. The forum began training partnerships between business corporations and NPOs. Thus, 1990 is often called the year of philanthropy renaissance (Imada 1999).

At this juncture, the earthquake hit Kobe city and adjoining towns in the Hyogo prefecture, which accelerated the process of civil society and voluntary activity in the affected area and gradually on to other parts of Japan. The extensive nature of voluntary efforts in this particular earthquake is attributed to: severity of the disaster; tremendous need among the victim population; intensive media reporting; and finally, the fact that the earthquake occurred during the local schools' winter break (Tierney and Goltz 1998). More than one million volunteers from different parts of Japan and abroad came to Kobe to serve the victims. Local governments were not prepared for this disaster, nor for the huge influx of volunteers. They had different types of relationships with these organizations: some city governments cooperated with these organizations effectively, some recognized them, but officially involved them in only limited capacities, and some even did not recognize them officially. Whether officially recognized or not by the local governments, the high spirits of volunteers were reported by the mass media extensively.

One year after the earthquake, Hanshin/Awaji Community Fund (HACF) was established in May 1996. The initial amount of the fund was 800 million yen (approximately US$8 million), which was supposed to be spent in 3 years by three different categories of grant: community redevelopment programs, philanthropic programs and NPO activities. In the same year (1996), the Japanese NPO Centre was established to support NPOs all over Japan. Nowadays, many similar organizations have been established to support the activities of smaller NPOs. After the earthquake, a new law was enacted in 1998: 'law to promote specified non-profit activities'. While victims and citizens in Kobe after the quake did not have time to join in lobbying for this law, they were very interested in lobbying for the financial assistance, which in turn lead to the enactment of another law: 'law to support disaster victims' in 1998. The spirit of civil-society activities, which gained high momentum in the immediate aftermath of the Kobe earthquake, continued to be strong in the environmental disasters, which followed, like the oil spill in the Sea of Japan and Tokyo Bay of 1997, where thousands of voluntary groups helped each other to mitigate the environmental degradation of the sea. Thus, there were two consequences of the Kobe earthquake: an emerging sense of self-governance, and stronger sense of community solidarity (Tatsuki and Hayashi 1999). These two elements brought new dimensions to civil society in Japan.

2.3 Post Kobe New Civil Society Model

Past disaster experiences show that immediately after the disaster, there are a "utopia of voluntary activities", which get settled as time passes. The next disaster strikes after the memory of past disaster events has faded. Cooperation with other sectors, especially to establish links between community organizations and professionals is an essential feature. Buchanan (1996) pointed out the difference between the roles of NGOs as outside agencies, and those from the perspective of community associations or membership associations. He also commented that although NGOs may see themselves as genuine partners of the local community and its organizations, in many cases the same view is not shared by the people. Community participation requires shared understanding between community members and the specialists from outside (including the NGOs and the disaster professionals) who aim to assist them (Twigg 1999). Inui (1996) provides examples from the Mano district in the affected areas of Kobe which exemplify the above factor in the form of a support network with the following activities: surveys of building safety, advice on buildings, building repair and cooperation in a professional joint reconstruction plan. The key issues for success are cooperation with the local community and an increase in the capability of the recipient community. As explained by Shaw and Goda (2004) with a case study in a small neighborhood of western Kobe, the formation of community organization [Danran] is an example of how to institutionalize the efforts at community level. The case study exemplifies how the community organization can generate its own resources through proactive participation in new community business. The problem of an ageing society may be unique in Japan, but the issues and processes of community involvement are universal, and can be applied elsewhere.

Based on the recovery experiences of Kobe, Shaw and Goda (2004) listed a few key lessons as follow: to incorporate the community into the decision-making process; to re-examine the values and traditions rooted in community and culture; third, to find the source of business opportunities in the community and thereby enable community-based organizations to sustain their efforts; and fourth, to promote community leaders to take action. A study of school children from different parts of Japan (including the affected areas of Kobe) suggest that while 70–80 % of them participate in disaster education and 60–70 % of them have high perceptions of earthquake risk, only 10–15 % of them actually take any action to reduce risk (Shaw et al. 2004). In the model of knowing, realizing and acting, there is always a gap between acquiring knowledge, perception and taking action. The results of the study also suggest that it is not the earthquake experience, but the community and family education, which are more effective in motivating a person to take risk-reducing action. Thus, future community participation should be aimed at teaching people effective risk-reduction measures.

The sustainability of civil-society action is another major factor, which needs careful attention. Figure 2.1 schematically shows the observed scheme of sustainable civil society, as evident from the experience of the Kobe earthquake (Shaw and Goda 2004). The residents' association supplies an important interface between community and administration with the help of professional and non-professional organizations like NGOs, NPOs and CBOs. Since this scheme generates its own resources within itself, the process is found to be sustainable. The association is also found to be effective in collective decision-making, and to represent the voice of the community. During the disaster events, this scheme is found to be useful, as exemplified by experience from other areas with different disasters in Japan, including flood and typhoon.

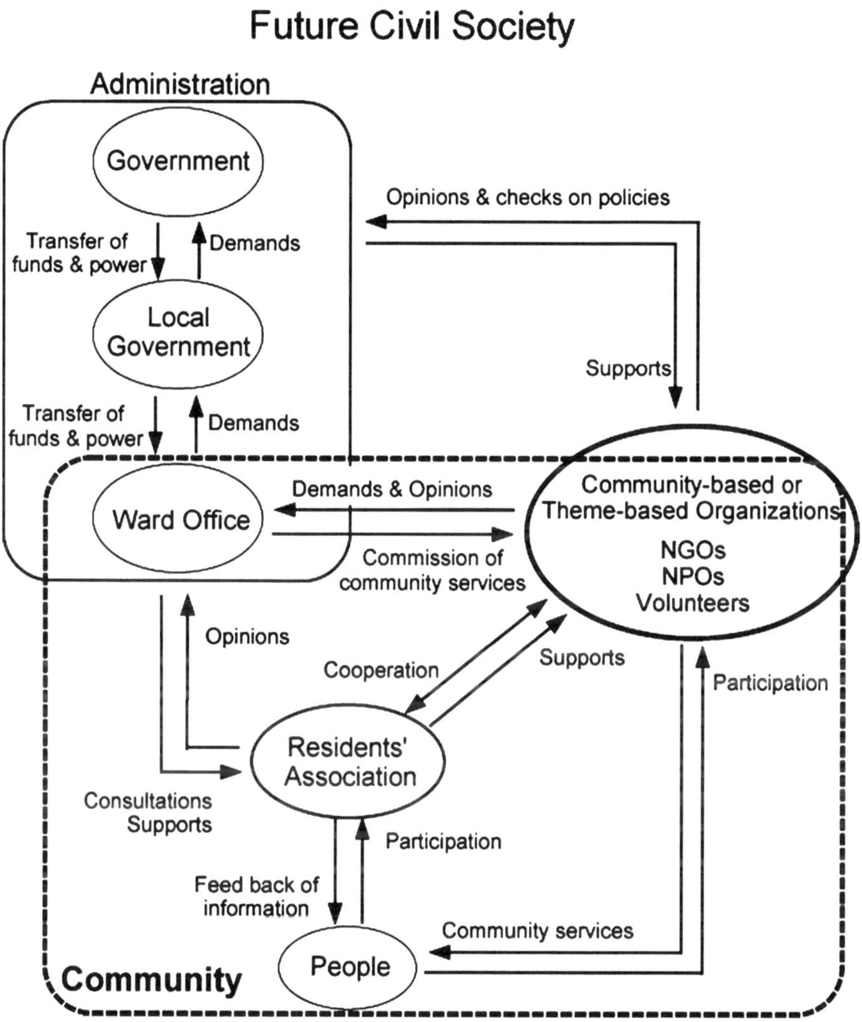

Fig. 2.1 Schematic diagram of future civil society as evolved from the recovery of Kobe earthquake

2.4 BOKOMI: New Sustainable Approach of Community Involvement

From the Hanshin Awaji-Earthquake, it is obvious that the age group of over 65 year-old is one of the most vulnerable groups in particular living in the density area with old houses. Learning from these findings, the Kobe City Government has been conducting an important initiative for developing "BOKOMI" with the aim to building resilience of its communities against disasters. BOKOMI is the short term of "Bousai Fukushi Community" [Disaster preparedness and welfare community], where the disaster preparedness is linked to daily welfare of the people. It was understood that the needs of the aged community is daily welfare. Therefore, to continue to the disaster preparedness activities, it is required to link this to welfare activities.

After the pilot phase in 11 districts, the BOKOMI concept was formalized in 1997 according to the Mayor's decision, and mainstreamed in all the school districts of Kobe City. BOKOMIs are established based on municipal elementary schools districts in Kobe City (Matsuoka et al. 2012). The total number of municipal elementary schools is 191. The number of BOKOMI steadily increased and reached 100 % coverage in 2008. The reason why BOKOMIs are based in elementary school districts is because 'welfare-community' groups were already established in each elementary should district and thus, disaster prevention activities could be integrated into these existing groups. In addition, elementary schools are designated as evacuation sites for communities in emergencies in Japan. These are the key reasons why BOKOMIs are established in each elementary school district.

The process of establishing BOKOMI in a district is based on multi-stakeholder consultation in the district. Firstly, the establishment of a BOKOMI is discussed and agreed by local government organizations, including the local city office (ward office) and fire station, together with leaders of local residents and other local multi-stakeholders. BOKOMI is a community-based organization comprised of local residents' associations, women's associations, elderly associations, child committee member, youth associations, PTA, local fire station, and local business entities. In order to support activities of BOKOMIs, the Kobe City Government provides various support measures like small funding, materials for community activities, rescue tools, training by fire professionals etc.

Main activities by BOKOMIs have two perspectives: disaster prevention and risk reduction activities and welfare related activities. These activities are combined and carried out together. Disaster-prevention and risk reduction activities by BOKOMI are as follow:

- Disaster drills and training
- DRR education program with schools
- BOKOMI junior team (fostering children's teams to lead and work on DRR activities)
- Public awareness event
- First-aid seminar, checking emergency materials and equipment
- Town watching and preparation of community safety map, risk reduction activities with rescue workers and fire fighters (identification evacuation root, removal of object blocking these roots, fixing furniture etc.)

Combining with welfare activity are as follow:

- Regular communication within communities to form their unity, so that they can take action, when emergency/disaster happens, considering needs of vulnerable groups such as elderly and disabled people.
- Learning how to support the people with special needs during disasters (elderly people and handicapped people).

2.5 Discussion

Immediately after the earthquake, most affected people were helped or rescued by friends, families and neighbors. Almost 60 % of residents were evacuated by their own efforts, and more than 30 % were rescued by the neighbors. This data indicates the importance of communities and neighbors in the immediate rescue operation. The main reasons local people are so effective in rescue activities are: information and knowledge of the community leadership within informal and formal community-based organizations; availability of small tools for rescue operation such as saws and crowbars.

After the rescue operation, the relief-and-rehabilitation phase began. In this, thousands of volunteers gathered from different parts of Japan. Different voluntary groups had coordination centers focused on different parts of affected areas. Needless to say, the prefecture, city and local governments had their coordination centers as well. In some places, there was cooperation with the NGO networks, in some places they acted independently. It is observed that while the role of outside volunteers gradually decreased in the temporary shelters, the local volunteer activities continued into the permanent housing phase. Changes were also noted in the nature of the volunteers. While non technical activities continued until the move to permanent housing, technical and networking work emerged at later stages of the reconstruction process.

The relief phase was followed by the reconstruction phase, in which government took the leadership. Roles of voluntary organizations became minimized in certain areas, which was attributed to: lack of sustained resources; lack of motivation to continue the efforts; changes in the organizational mandates (some organizations focused on rescue and relief only); and lack of technical skills to contribute to the reconstruction process. Among different activities, the creation of temporary shelters, identifying special zoning areas, restoring lifelines and infrastructures were the priority issues. From the government perspective, the reconstruction phase lasted for 3–5 years, until the housing and infrastructures were fully reconstructed. However, according to the people's perspective, the reconstruction phase it continued for several years.

Kobe earthquake has changed the concept of disaster risk reduction in Japan. Earlier, the role of civil society or the NGOs in Japan was to be a watchdog in government activities. However, the recovery process has shown it is the government-NGO collaboration, which can make a successful and sustainable recovery. No government, either in developed or developing country, can do risk reduction by itself. It needs collaboration, cooperation from different sectors. In past Kobe, the emergence of new civil society is a very important movement, Japan has experienced. The civil society played a very important role to bridge the gap of government and community. As explained earlier, the CSOs also played important role with the local resident organizations to provide them different types of help and support from different professional expertise. The local governments started new schemes to provide financial supports to the CSOs and NGOs to conduct professional work along with the local communities. This is considered as a new development in redefining the roles of CSOs and NGOs in Japan, where professionalism came quite strongly. A combination of "humanitarian heart" and "development brain" was the ideal future of CSOs and NGOs working in the disaster related field.

The volunteer coordination system also developed very strongly after the Kobe disaster. While, in post Kobe there were different levels of confusion and lack of information flow and lack of coordination, bringing these lessons further, nation wide volunteer coordination center was established. A volunteer roster and registration system was developed, where people can register themselves as potential volunteers before a disaster happens. The volunteer database is shared closely with the local governments, and a coordination mechanism was developed to dispatch the right volunteer to the right place based on the local needs. Thus, Kobe earthquake is considered to make different changes in community work, volunteer work, role of CSOs and NGOs etc. Much of these efforts showed positive response in different other post Kobe disaster events in Japan.

Acknowledgements This chapter is based on different earlier work over more than 10 years, which is highly acknowledged.

References

Buchanan A (1996) Partnership and accountability. Appropriate Technology 22(4):6–8
Deguchi M (1999) A comparative view of civil society: Japan–America civil society; Program 1: civil society in Japan and America: coping with changes. Washington, DC
Imada M (1999) Transition to civil society in Japan: Japan–America civil society, program 4: civil society: where the private, public and press meet, Washington
Inui K (1996) The Role of volunteers with professional skills in the reconstruction of urban areas. In: Goater JF (ed) Innovative urban community development and disaster management. UNCRD, Nagoya, Japan
Kinmokusei International Project (1999) Key terminology in restoration from the Hanshin earthquake disaster. Kinmokusei, Kobe
Maskrey A (1989) Disaster mitigation: a community-based approach. Oxfam, Oxford
Matsubara A, Todoroki H (2003) Japan's 'culture of giving' and nonprofit organizations: NPO Web. Available at: http://www.npoweb.gr.jp/english/cgp.html
Matsuoka Y, Joerin J, Shaw R, Takeuchi Y (2012) Partnership between city government and community-based disaster prevention organizations in Kobe, Japan in community based disaster risk reduction. Emerald, Bingley, pp 151–184
Miyamoto K (1996) Problems with the reconstruction plan for the Hanshin-Awaji area. In: Goater JF (ed) Innovative urban community development and disaster management. UNCRD, Nagoya
Nakagawa Y (2003) Disaster and development, applying social capital in disaster recovery. Master's Thesis. Kobe University, Kobe
Nakagawa Y, Shaw R (2004) Social capital: a missing link to disaster recovery. Int J Mass Emerg Disasters 22(1):5–34
Noguchi Y (1995) The 1940 System: A farewell to war-time economy (in Japanese). Toyokeizaishinpo, Tokyo
Raphael B (1986) When disaster strikes: a handbook for the caring professions. Hutchinson, London

Shaw R, Goda K (2004) From disaster to sustainable community planning and development: the Kobe experiences. Disaster 28(1):16–40

Shaw R, Shiwaku K, Kobayashi H, Kobayashi M (2004) Linking experience, knowledge, perception and earthquake preparedness. Disaster Prev Manag 13(1):39–49

Tatsuki S, Hayashi H (1999) Determinants of the changes in residence and life reconstruction among the 1995 Kobe earthquake victims. In: The 24th natural hazard workshop, Boulder

Tierney K, Goltz JD (1998) Emergency response: lessons learned from the Kobe earthquake. See http://www.udel.edu/DRC/preliminary/260.pdf

Twigg J (1999) The age of accountability? Future community involvement in disaster reduction. Aust J Emerg Manage 14(4):51–58

Chapter 3
Typhoon, Flood and Landslide Related Disasters in Japan: Role of Communities

Rajib Shaw

Abstract Japan, due to its critical location is prone to typhoon, and related disasters like floods and landslides. There has been a strong history of risk reduction in Japan with community approaches. This chapter demonstrates the evolution of community roles in typhoon, flood and landslide related disasters over time with four specific examples: 1999 Hiroshima sediment disasters, 2004 typhoon disasters in Saijo City and Toyooka city, 2009 and 2011 rainfall disasters in Shiso city. These examples elaborate different issues like: evolution of community hazard maps and community participation in the drills (in Hiroshima), implementing town watching in the schools to enhance school community linkages (in Saijo), enhancing the roles of volunteers in post disaster debris management (in Toyooka), day time and night time disaster drills and community hazard map making (in Shiso). Collectively, it is shown that community plays in important role in different aspects of risk reduction activities before, during and after disasters.

Keywords Community hazard maps • Disaster drills • Sediment disaster • Town watching • Volunteer management

3.1 Introduction

Mountains and hilly areas comprise about 70 % of the land area of Japan. Topographical and geological factors, such as steep mountains, fast-flowing rivers, and unstable and soft ground, combined with weather conditions like typhoons and localized torrential rain climate and frequent earthquakes contribute to making

R. Shaw (✉)
Graduate School of Global Environmental Studies, Kyoto University,
Sakyo-ku, Yoshida-Honmachi, 606-8501 Kyoto, Japan
e-mail: shaw.rajib.5u@kyoto-u.ac.jp

R. Shaw (ed.), *Community Practices for Disaster Risk Reduction in Japan*,
Disaster Risk Reduction, DOI 10.1007/978-4-431-54246-9_3, © Springer Japan 2014

Table 3.1 Sediment disasters in Japan

Date	Location	Occurrence
January 1986	Nou Town, Niigata Prefecture	*Avalanche caused by heavy snowfall*: The large scale surface layer avalanche in Tanaguchi district crushed 11 houses, killed 13 and severely or slightly injured 9 people
July 1990	Ichinomiya Town, Kumamoto Prefecture	*Debris flow accompanied by trees*: Debris flow caused by heavy rain and volcanic ash ejected from Mt. Aso's eruptions killed 8 people and destroyed 151 houses totally or partially
June 1991	Shimabara City, Nagasaki Prefecture	*Pyroclastic and debris flow from Mt. Unzen eruption*: Serious damage was caused by pyroclastic and debris flows from the eruption of Mt. Unzen and killed 43 people at Fugendake
January 1995	Nishinomiya City, Hyogo Prefecture	*Landslide caused by earthquake*: A landslide occurred on the left mountainside over an existing dam and with a residential area situated close to the collapsed sediment of 1,500 m^3
August 1998	Nishigo Village, Fukushima Prefecture	*Sediment disaster caused by a localized torrential downpour*: Localized torrential rainfall (total rainfall: 1,267 mm; maximum rainfall per hour: 90 mm) brought about earth and rock flowing into a social welfare facility killing five people
June 1999	Kure City, Hiroshima Prefecture	*Slope failure disaster caused by localized torrential rainfall*: Debris flows and slope failures caused by localized torrential rainfall under the influence of the Baiu front killed one person and destroyed six houses totally and partially
March 2000	Abuta Town, Hokkaido	*Mudflows caused by Mt. Usu eruption*: An enormous amount of heated mudflow flowed down into the Nishiyama River running through the Toyako spa resort, inflicting damages and carrying away bridges on national and town roads

Source: Japan Sabo Association (2001)

Japan highly prone to sediment disasters. As a consequence, debris flows, landslides and slope failures occur throughout the country every year (Infrastructure Development Institute-Japan and Japan River Association 2006). Approximately 1,000 sediment disasters occur in Japan every year due to the existing natural conditions (Uchida et al. 2009) and account for about a half of the dead and missing by natural disasters (excluding those by the Southern Hyogo Earthquake, January 1995) (Japan Sabo Association 2001). The occurrences of sediment disasters in 1986–2000 are described in Table 3.1.

To protect lives and properties from sediment disasters, structural measures have been undertaken by building facilities and infra structures (Table 3.2) as well as non-structural measures through a system of warning and evacuation and restricting and controlling new residential land development in areas vulnerable to sediment disasters (ibid.) In addition, several laws were enacted to promote structural preventive measures such as the "Sabo Law", "Landslide Prevention Law", and "Law Concerning the Prevention of Disasters due to the Collapse of Steep Slope" and the "Act on Sediment Disaster Countermeasures" to promote nonstructural preventive measures.

Table 3.2 Brief description of structural preventive measures

Preventive measure	Description
Sabo for volcanoes	Protects people from pyroclastic, debris and lava flows associated with volcanic activities
Slope failure preventive works	Protects people from disasters caused by a slope failure
Debris flow preventive works	A Sabo dam with a sand pool space is built where a debris flow runs down and is deposited so as to directly receive and capture a debris flow
Measures against wood debris	Silt type dams are built to effectively check the run-off of trees fallen by wind or other causes
Avalanche prevention works	Protective fences are installed to protect local communities from snow avalanches
Landslide prevention works	Groundwater is drained and piles are driven into the ground to prevent the ground from starting to move and sliding downward
Sabo dam	Sabo dams control a large amount of sediment outflow like a debris flow without causing damages along the downstream
Hillside works	Vegetation works, land retainers and drainage systems are installed on mountain and hill sides to restore vegetation and prevent slope failure and sediment run-off
Consolidation works/ revetment works	The works create a space that serves to prevent erosion of river and valley banks and allow people to appreciate waterfront landscape. Streams are provided with fish passes to conserve the ecosystem

Source: Japan Sabo Association (2001)

In this chapter, several examples of typhoon, flood and sediment related disasters are mentioned. First, the aftermath of Hiroshima sediment disaster is discussed with specific implications to development of community hazard maps. Second, 2004 typhoon disaster and the initiation of school based community disaster risk reduction are analyzed for the city of Saijo. Third, the volunteer activities in 2004 typhoon disaster in Toyooka are discussed. Finally, 2009 and 2011 typhoon and catastrophic rainfall events in Shiso city in Hyogo Prefecture are analyzed with implications to preparation of my community hazard map and participation in disaster drill. Finally, the experiences of community involvement are analyzed for these water related disasters.

3.2 Hiroshima Sediment Disaster and Aftermath: Implication of Community Hazard Maps

Hiroshima City is the largest city in the Chugoku region. Between the 1970s and mid-1990s, the city saw steady population growth as people from different places came in to settle. As a result of this growth, the city has spread out even over the mountainous areas. As a consequence, Hiroshima City has the highest risk of sediment disaster in Japan. From their past experiences and memory, longtime residents know where the risky places are and when a disaster might possibly happen. But new residents do not have this knowledge. Moreover, some wards of the city show symptoms of an aging population, increasing the city's vulnerability to disasters. Another issue is the geology of some areas, which makes them susceptible to sediment disasters (Takeuchi and Shaw 2009a, b). Granite, which becomes softened and fragile when soaked in rainwater, is widely distributed and it becomes a contributing factor to sediment disasters.

On June 29, 1999, due to an atmospheric depression, there was heavy rainfall (over 150 mm in 3 h) in Hiroshima City. Prior to this event, similar heavy rainfall occurred twice in the same region (June 23–24 and June 26–27). Disaster damage included debris flow at 139 sites; the collapse of 186 buildings; a death toll of 31, with 1 person missing; and 154 houses damaged. Hiroshima City suffered heavy damage, especially in the new residential areas of Asaminami Ward and Saeki Ward. After the disaster, Hiroshima Prefecture and Hiroshima City adopted various soft and hard countermeasures (Takeuchi and Shaw 2009a).

There was 70–100 mm of rain twice in Hiroshima on June 23–24 and 26–27, 1999. On June 29, 1999, a stationary rain front caused torrential rain in Hiroshima and an avalanche of rocks and earth affected several towns. On the day of the disaster, there was an official announcement at 10 o'clock to inform residents through TV and radio and through the community network.

The 29 June 1999 disaster demonstrated how the expansion of residential areas aggravates vulnerability to sediment disasters of certain locations. The national government, thus, established the Act on Sediment Disaster Countermeasures, which became effective in 2001. A chronology of events leading to its enactment is shown in Table 3.3. Essentially, the Act on Sediment Disaster Countermeasures seeks to institute comprehensive non-structural measures to protect people from sediment disasters. Prefectural governments, in particular, are tasked to undertake non-structural measures including public disclosure of risk information of areas prone to sediment disaster and the development of warning and evacuation system, restriction on new land development for housing and other specific purposes, and promotion of relocation of existing houses in these areas (Infrastructure Development Institute-Japan and Japan River Association 2006). Among sediment disasters, the act specifically targets debris flow, landslide and slope failure. The act proposes some steps in its implementation such as a basic survey based on the guidelines for sediment disasters prepared by the Ministry of Land, Infrastructure, Transport and Tourism; designation of disaster warning areas; and development of early warning and evacuation system.

In disseminating sediment disaster information, hazard maps are distributed to local citizens to call attention to sediment disasters. A "Hazard Area Map on Sediment Disaster" shows the relation between sediment disaster hazard spots (i.e. debris-flow hazard torrents, slope-failure hazard spots and landslide hazard spots), evacuation places and evacuation routes. Furthermore, information on sediment-related disaster warning areas is given by direct mail to people living in the area, which is prone to a sediment disaster. Together with the "Hazard area map of sediment disaster" and "Municipal disaster prevention plan", the mail calls for the need for precaution and evacuation in case of heavy rain (Japan Sabo Association 2001).

The Act on Sediment Disaster Countermeasures was revised in July 2005 to make compulsory the mapping of "Sediment-related Disaster Special Warning Areas" and distribution of such hazard maps. At the same time, the Guidelines and

Table 3.3 Chronology of events leading to the enactment of the act on sediment disaster countermeasures

Year/date	Event
1999	
June 29	Hiroshima disaster (landslides of 325 places; 24 lives lost)
July 8	The Ministry of Construction, Disaster Prevention Land Management Promotion Office was started. It was determined to organize a "project team for comprehensive sediment disaster prevention"
2000	
February 4	The River Planning Council made recommendations with respect to "legal institutions for comprehensive sediment disaster prevention"
March 14	Bill for "Act on Sediment Disaster Countermeasures" was approved by the Cabinet Council (submitted for deliberation to Congress)
April 18	The bill was unanimously passed in the House of Councilors
April 27	The bill was unanimously passed in the House of Representatives
May 8	"Act on Sediment Disaster Countermeasures" was promulgated (2000 Law No. 57)
2001	
March 28	Cabinet order for the "Act on Sediment Disaster Countermeasures" was promulgated (2001 Cabinet Order No. 84); Cabinet Order for "Revision of Part of Construction Standard Law" was promulgated (2001 Cabinet Order No. 85)
March 30	Ministerial Ordinance for the "Act on Sediment Disaster Countermeasures" was established (2001 Land, Infrastructure and Transport Ministry Ordinance No. 71)
April 1	The "Act on Sediment Disaster Countermeasures" was enacted
July 9	Basic Guidelines for the "Act on Sediment Disaster Countermeasures" was established (2001 Land, Infrastructure and Transport Ministry Notice No. 1119)

Source: Japan Sabo Association (2001)

Exposition for Sediment Disaster Hazard Mapping was prepared by the Erosion and Sediment Control Division of the Ministry of Land, Infrastructure, Transport and Tourism. The guidelines indicate that maps must include information stipulated by the Act on Sediment Disaster Countermeasures such as (1) the means for transmitting information on sediment disasters, (2) evacuation locations in the event of debris flow and steep slope failure, and (3) other information required to ensure smooth precautionary evacuation (Uchida et al. 2009). These hazard maps must be produced by all heads of municipalities who must also ensure the smooth precautionary evacuation of residents through the dissemination of information using appropriate means. In the aftermath of the disaster, the Hiroshima provincial and city governments took an innovative approach to involve local residents in implementing the "Act on Sediment Disasters countermeasures."

A study by Umakoshi (2011) attempted to throw light on this important issue of resident participation through a detailed questionnaire survey, field visit and hearing survey in Asaminami ward of Hiroshima City. The study area was characterized by a mixture of old and new residential areas on the mountain slopes, and experienced major landslide and flooding several times in the past, including the catastrophic sediment disaster of 1999. It reveals that there was a three step process of implementation of the "Act on Sediment Disasters countermeasures" in the target area: Step 1—setting of sediment disaster warning area, and explanation to the local resident association and residents (September–October 2007), Step 2—development of evacuation system, prepare evacuation manual with the support from the ward office and fire department (November 2008), and Step 3—implementation of evacuation drill based on the evacuation manual (January 2009).

One of the key challenges of the sediment disaster is timely warning, and people's evacuation. Literature review conducted suggests that there are several factors which affect the evacuation behavior of people and communities, some of which are psychological, some related to hazard characteristics and some related to demographic factors like age, gender of the people. It is also found that the regular evacuation drill enhances the effectiveness of actions during emergency situation. Public education and awareness-raising play certain roles in understanding the hazard, vulnerability and risk of the region.

The resident perception survey in the Asaminami ward of Hiroshima City shows that there is a high interest in the residents to know disaster related issues, and almost half (51 %) of the total respondents (total is 385) points out that they are aware of the evacuation route, and 41 % have the evacuation manual with them. Sixty-three percent responded that they have good relation with the neighbors. Fifty percent of the respondents have experience of the 1999 sediment disaster, however, only 12 % people evacuated in this disaster. Almost 60 % respondent was aware of the "Act on Sediment Disasters countermeasures," and people in the red zone (high risk areas designated by the Act) are aware of the risk of the area. Almost 805 of the people from the red zone areas are aware of the evacuation manual, but only 12–13 % people actually participated in the development process of the manual. However, 50 % have the manual, and more than 32 % participated in the drill conducted in January 2009 (Umakoshi 2011).

All the above observations point out interesting relationships between the disaster experiences, houses belonging to the designated areas (red or yellow), participation in the manual development and drill, and disaster awareness. The location of the study area is of prime importance in this regard. Unlike other remote mountain villages or neighborhoods, the current study area is located very close to the city center, and is well connected to the center through public transport. Therefore, the area is characterized as a residential area. The areas being an urban one, people have limited time to afford for the community activities. The results show that although there was a strong government intervention to increase resident's participation in the manual development and Act implementation process, very few people actually participate in the activities.

This is possibly a typical character of the urban areas, and the study location is no different in this aspect. The cross correlation shows that the people who participate in regular community activities (organized by residential association, called *Jichikai* in Japanese), have higher levels of awareness about the designated areas, and they also participate actively in the manual development as well as the evacuation drills. People's network become stronger through community events like local festivals, and the people who are living in the regional for a longer period of time or those who have their own houses, are more aware of the risk of their houses.

It is required to think how the governments can facilitate the participation of the residents in the different disaster related activities, including the manuals and drills.

In the urban context, possibly more internet or IT (information technology) based approaches can break the boundary of people's participation. Different social networking schemes can be used for this purpose. Roles of facilitators become important in urban context, and the local government needs to develop a strategy to enhance identification and deployment of the facilitators in the local communities. Finally, evacuation drill is a practice and simulation exercise. This is important for the redundancy. Although there is no visible impacts on evacuation drills immediately, but by conducting the drill, the redundancy in the system is enhanced which become effective even for the warning in short notice.

Finally, the evacuation manual cannot be a static document. The key point of the manual is its updating, monitoring and evaluation. Although people may not have participated in the first process of the manual development or the evacuation drills, but their participation can be enhanced in the monitoring and evaluation process through multi-stakeholder cooperation. The local institutions like schools, health centers, business communities or local community houses can play different roles in revitalizing the local communities and to encourage them in participation process. Hiroshima City is actively involved in the public awareness and education programs through different pamphlets, brochures and internet base information system through websites. This needs to be continued, and roles of media (like newspaper, TV, radio) become increasingly important in public education.

Thus, for the implementation of the "Act on Sediment Disasters countermeasures," people participation can be ensured through education, awareness, innovations in social networking, and providing information through all different types of media. Community participation is a time consuming issue, and this can only be resolved over time through gradual engaging process.

3.3 Typhoon Disaster in Saijo: Role of School Education

The city of Saijo (with population of 116,000 as of 2004) in Ehime Prefecture was affected by a series of typhoon in 2004. The city is classified broadly into four: plain area along the coast; hilly area between Saijo City and Nihama City; hilly terrain ranging in the north side of the median tectonic line along the south side of the plain; and precipitous mountains in the southern side of the median tectonic line. There is Mt. Ishiduchi, the highest mountain in the western part of Japan. Two big rivers, Kamo River and Nakayama River, flow in the center of the city. Saijo City is famous for its spring water called "Uchinuki." There is a traditional annual festival called "Saijo Matsuri" in October when almost all people, both young and old, get wildly excited. Each "Jichikai" (resident association) has its own "Danjiri" (portable shrine) and the relationship among people in the communities is strongly felt. A "Jichikai" is a neighborhood association which is organized in each area in the municipality at the people's own initiative. There are 540 Jichikai in Saijo City and they make 28 Jichikai unions (Yoshida et al. 2009).

At the time of the typhoon no. 21 and 23 in 2004, mountainous area of Saijo City was especially damaged. Land condition and concentrated heavy rain were the major factors, but there were other reasons concerning concentrated aged population in the region. So, some elderly people had difficulty in evacuating and needed help of young people. Low awareness of disaster prevention is also a problem. Plain area is rather urban in nature, and there are many young people. So, it is necessary to make "disaster prevention network" between the plain area and the mountainous area, so as to help elderly people in the mountainous area in case of a disaster. As the driftwood stuck with bridge pier caused flood to the plain area, disaster in the mountainous area have bearings with that in the plain area. Both residents have to know each other about the circumstances (Yoshida et al. 2009).

After the disaster, Kyoto University started a program with Saijo City education board to enhance the linkage of school and communities through participatory neighborhood watching. For these reasons, mountain watching was implemented in Saijo City. Mountain watching is just like town watching and it is conducted in the mountainous area. Main target was children, and also residents in the mountain, teachers, municipal officials and forest workers were involved. The working field was upper area of a river along school. Participants watch the site damaged by the typhoon in 2004 and hear the story from victims.

At the same time, town watching was implemented in the plain area. The main target was students and teachers, parents, Jichikai and municipal officers. They walked around the school zone and search for dangerous places, useful facilities in case of disasters and favorite places, which they don't notice otherwise in daily life. At first, town watching was implemented in five elementary schools and mountain watching in three junior high schools as "disaster education program", which was an activity of 12-year-old education project. The project started in 2005, and continued till 2007. From 2008 onward, the city education board took leading role in continuing the process in the implemented schools, and also expanded it to all primary and secondary schools in the city.

One of the key elements of the education process was the school community relation. The local residents often have lots of local information (not only related to disasters, but also local culture, values, tradition). The key focus of the town watching was to engage the students with the communities to dig this local information and bring it to the education process. Teacher played the role of facilitator in this process. The involvement of the education board was important to sustain and institutionalize this effort.

3.4 Typhoon Disaster in Toyooka: Role of Volunteers

On Wednesday 20 October 2004, the typhoon 23 (the same typhoon which hit the city of Saijo), Japan's severest storm in many years, swept through most of the southern half of Japan. People were overcome by the massive waves and flash floods triggered by the typhoon's heavy rains and strong winds, which left at least 93 people dead, and some 490 injured, 97 of them severely. The number of casualties from the typhoon was the highest from a typhoon since October 1979 when 115 people were killed or presumed dead. Many of the deaths occurred when homes were buried in landslides caused by 2 days of torrential rain. Television images showed rescue workers digging through the remains of wooden homes for survivors.

Compared to typhoons that usually strike Japan, typhoon 23 had a number of characteristics that increased its impact, including strong wind speed and gradually increasing energy, deflection from usual typhoon paths, and the fact that the hardest hit areas were affected during the night. The combination of these factors enhanced the impact of the typhoon on people, infrastructure, leaving small cities particularly vulnerable. The number of typhoon related casualties is the highest in over a quarter of a century, and it further destroyed homes, damaged and flooded many residences. Storms and floods killed over 100 people in Japan in 2004, and the damage from this typhoon season already exceeds the bill for weather-related damage in 2003, which amounted to 940 billion yen (about US$9 billion).

Toyooka is a small city, with a population of 48,000. It has a river running from south to north, a wide-open rice field in the middle, surrounded by mountains on all sides, and a narrow passage to the sea to the north. It was traditionally a farming community, with recent development making it a commercially important city in the region. It is famous as a nesting place for migrating White Oriental Storks—one of the last remaining sites in Japan. As a result, tourism is also an important element of the city's economy.

In the aftermath of the typhoon, volunteers from different parts of Japan arrived in Toyooka city, and helped in the immediate response, especially for cleaning up typhoon waste. The local government noted that a typhoon's aftermath is different from that of an earthquake, since the first priority after a typhoon is to clean the residences, and remove destroyed and damaged household materials. For this, management of volunteers (for example, ensuring that volunteers were distributed to all needy areas of the city) was critical, and this was effectively carried out by nonprofit organizations (NGOs). After any disaster, psychological/mental care of the affected people is of utmost importance. In the aftermath of a typhoon, several programs were carried out by volunteer groups, local governments and other organizations to cheer up residents, children and aged people. These included entertainment programs, shows, and sending flowers, toys, and pets to the children.

After the Kobe earthquake of 1995, there were several issues on volunteer activities: lack to preparedness of volunteers themselves, coordination of government and volunteer coordination center, lack of information, leading to mismatch of need and dispatch of volunteers etc. All these lessons were useful to make coordination of volunteer activities in Toyooka. The volunteer coordination group from Kobe was effective in setting up the coordination center inside the city office, which enhanced smooth information flow with the local government and local residents.

The volunteers were very active in organizing the cleaning activities after the disaster, which was crucial for the typhoon disasters. A significant amount of waste was produced, which is equivalent to the almost 3 years of waste production of the city. One of the crucial issues of disaster waste is the domestic waste, which is generated after the disaster. Depending on the nature of the disaster, it has its characteristic features. For the earthquake, it is rather the dry debris, mainly from the building materials. The electronic items sometimes get crushed under the building rubbles, and become unusable. For the typhoon related disasters, the key point is flooding and flood-induced mud and related debris which is transported and deposited in the houses. A traditional Japanese house has tatami mat in the first floor, and most of the electrical appliances like washing machines, refrigerator, television etc. are kept in the first floor. Thus, when the first floor is flooded, all these electronic items get affected. It is not that all the items become unusable; however, this is mostly psychological issues that people do not want to use the appliances which are affected by the dirty water and mud. Thus, people throughout these electrical appliances, which cause a huge burden to the local government's limited resources. When these waste materials are thrown out, volunteers can play an important role. A study by Satoh (2007) points out that if these materials are segregated in the initial disposal areas (which are usually outside the house, on the road side), the treatment cost becomes almost one third of that when these are mixed. However, due to psychological issues, the affected people do not have that much time and mental situation to segregate these wastes at disposal time. The study also showed that the city government who undertook a simple volunteer training program (with some awareness raising leaflet and pamphlet for segregation at source) has been able to reduce the cost of debris management. The Tokooka experience is one of the good examples of volunteer involvement in debris management and segregation, which is replicated in other disasters later, including the East Japan Earthquake and Tsunami of 2011.

3.5 Catastrophic Rainfall Disaster in Hyogo: Role of Disaster Drills and Disaster Prevention Maps

Shiso city is located in the west part of Hyogo Prefecture. It has an area of 658.60 km^2 with a population of 40,945 as of 2011. Shiso city is strategically situated at the middle of Ibo River, which has a basin area of 810 km^2 and a length of 70 km. Most of Shiso city is covered with mountain forest and flat land is very little. The residence area is thus developed along the flood plains of Ibo River and Hikihara River. The major industries in Shiso city are timber, timber product and furniture. Shiso city is also well-known for the manufacture of Somen, a kind of Japanese noodle, due to the fresh and good quality of water from the Ibo River used during the production process. However, at present, these industries are declining due to reduced and aging population.

Magari is a district in Ichinomiya town of Shiso city. It is located in the confluence of Ibo River and Hikihara River and has 114 households. The water level of the river tends to increase rapidly due to the narrow river width and steep slope, which is about 1/190 in the district. On August 8–11, 2009, several areas from Kyushyu to Tohoku received heavy rains from typhoon. Rainfall exceeded 700 mm in Shikoku. In terms of daily rainfall, rainfall exceeded 327 mm in Sayo town of Hyogo Prefecture. In Shiso city, rain started around midnight of the August 9th. In Kobe district of Ichinomiya town, daily rainfall amounting to 191 mm was observed while hourly rainfall of 63 mm was observed in Mikata district of Ichinomiya town. As a result, evacuation was ordered in Magari district because the water level of Ibo River increased to 3 m, which is the flood alert level set by MLIT (Ministry of Land, Infrastructure and Transport). This rainfall corresponds to similar rainfall in 1976 and 1990. Magari district suffered extensive flood damage due to the rain. Seventeen houses were inundated above floor level while 39 houses were inundated under floor level due to the flood.

At the aftermath of the disaster, the Magari district started to develop "My BOSAI MAP." "My Bousai Map" is a unique exercise to develop the local disaster prevention map, which is customization of city/prefecture disaster prevention map with local information on past disasters. The development process consists of seven steps including (1) Planning schedule of making My BOSAI MAP, (2) Workshop for making My BOSAI MAP, (3) Town watching exercise, (4) Distribution of the map, (5) Disaster drill, (6) Validation of effects and issues in the practice and (7) Improvement of issues and arranging technical know-how. This flow is based on the PDCA cycle, a management cycle for action, suggesting a "Plan", "Do", "Check" and "Action" flow of action. In this process, the MLIT and local government disaster prevention officials played a facilitator role to engage the local communities and provide some technical suggestions and help.

MLIT conducted a questionnaire survey to validate the effects and issues of the pilot program after the "Town watching exercise", "Distributing My BOSAI MAP" and "Disaster drill". In answer to the question (for Town watching) "Do you think town watching exercise is needed for making My BOSAI MAP?", the participants who answered "Yes". Sixty-eight percent of the participants think that town watching exercise is needed to draw the My BOSAI MAP because it is possible to check the dangerous areas and points in the district through the exercise. In addition, 66 % of the participants think town watching exercise is needed because it is possible to check the dangerous area and points in the district which was not known before now. Next, the participants were asked about the strategy to conduct a town watching exercise effectively. Sixty-three percent of the participants think two times town watching (Noon and Night) is needed for effective exercise. In addition, 49 % of the participants think participation of a wider age group is needed and 44 % think participation of Fire brigade is needed. This is a quite significant step that in Magari district, town watching was conducted in the day-time as well as in the night time. The issues and problems observed were different, because many of the residents are aged, and have physical issues related to the evacuation during night-time, without some aid. For the question regarding the "distribution of my Bousai map", 93 % of the respondents think it is possible to understand the dangerous area and points through the map. However, some respondents think it is difficult to understand the information because the contents of the map are complex. Also, 10 % of the respondent's answer that "I have known all dangerous area and point in the map before checking the map", 31 % of the respondents answer that "I have not known the dangerous area and point before checking the map". Fifty-nine percent of the respondents answer that I have not known some dangerous area and points before checking the map. In the question related to the "disaster drills," the questions are "What contents in the map did you feel easy- to-understand at the disaster drill," 69 % of the participants answer "Evacuation route" and 62 % of the participants answer "Flood risk area".

The same area was affected by heavy rain in 2011. The rain occurred in 10th and 11th of May 2011. The event was not as serious as the one in 2009. The evacuation instruction was announced in Magari district at 20:50 on 11th of May because the water level of Ibo River increased to flood attention level. However, flood did not happen because rain stopped and gradually water level decreased. A questionnaire survey (total respondents: 106) was done in the same area on the effectiveness of earlier preparedness (after 2009 disaster) to the evacuation behavior of 2011 rainfall (Suda 2012). Seventy-six percent of the respondents evacuated during the heavy rain. On the reason, 73 % of the respondents answered, "I evacuated because I knew the evacuation instructions", 43 % answered "I was urged to evacuate by fire company or head of neighborhood association", and 31 % answered "Because I felt danger due to lessons from the heavy rain in 2009". However, the respondents are few who answer, "Because I know the risk in this area through the My BOSAI MAP and town watching exercise". On the other hand, the reasons why people didn't evacuate are that "Because I think stay-home is safe" (33 %) and "Because I did not feel the rain has disaster risk" (29 %). The reasons why respondents think staying home is safe are that "Because evacuation route is not safe" and "I think that I do not need to evacuate because water level was under the indication mark when I checked it". Thirty-nine percent of the respondents use the My BOSAI MAP during the heavy rain on May 2011. Seventy-one percent of the respondents who used the My BOSAI MAP answer, that "The map was effective for evacuation". The reasons why My BOSAI MAP is effective for evacuation are that 90 % of the respondents answer, "I could check the evacuation site", 51 % answered "I could check the evacuation route".

This experience shows the importance of developing the community's own disaster prevention map. This is a participatory process, and needs to be considered as a dynamic approach. The disaster prevention map needs to be updated regularly with local information. The disaster drills both is day-time and night-time provides important information, which needs to be incorporated in the local disaster prevention map. Local residents have lots of local information, which needs to be linked to the expert knowledge to develop the local disaster prevention map. Testing, evaluation and updating of the disaster prevention map is important.

3.6 Analysis and Way Forward

Four examples of community involvement and community practices for disaster risk reduction are provided in this chapter: 1999 Hiroshima disaster, 2004 typhoon disaster in Saijo, 2004 typhoon disaster in Toyooka and 2009, 2011 typhoon and rainfall disaster in Shiso. Table 3.4 summarizes the key observations from these four examples.

Table 3.4 Summary of key observations in case studies

City/Province	Year	Nature of disaster	Key observations
Hiroshima City, Hiroshima Prefecture	1999	Sediment disasters	– Change of national policy – Involvement of local residents in disaster prevention map and regular drills, development of self manual
Saijo City, Ehime Prefecture	2004	Typhoon disaster	– Involvement of local schools for town watching process – Participatory disaster education
Toyooka city, Hyogo Prefecture	2004	Typhoon disaster	– Volunteer involvement in debris segregation and clearance – Community and local government cooperation
Shiso city, Hyogo Prefecture	2009 and 2011	Typhoon and catastrophic rainfall	– Local residents own town watching [day time and night time] – Development of community's own disaster prevention map – Regular disaster drills

Japan is located on the eastern edge of monsoonal Asia and its climate varies according to the seasonal and regional conditions. Typically, heavy rains hit various parts of the country both during the rainy season in June and July and during the typhoon season from August to October. This precipitation is predominantly in the form of locally specific temporary downpours. In winter, the northern part of the country usually receives heavy snowfall, which causes prolonged snowmelt floods in spring. The average annual precipitation in Japan is rather high, around 1,500–1,600 mm. Precipitation in Tokyo is twice as much as other large cities in western countries. Some 50–60 % of the annual precipitation in the Pacific coast of Japan is concentrated in the months from June to October. Thus, typhoon, typhoon induced rainfall and rainfall induced landslide and sediment disasters are rather common in Japan.

Japan's modern engineering application started in 1868 during Meiji period, when the country incorporated river improvement works, first for navigation and then flood control. The first River Law was enacted in 1896 reflecting the occurrence of a series of tragic floods in 1880s. It declared the public ownership of rivers and river water and assigned to the central government the flood control responsibility for works that demand large funds, high technology and nationwide planning while the local governments assumed the principal responsibility of ordinary flood management. The Law authorized water used in rice paddy irrigation as a historically vested right. The basic strategy for flood control was to lead water to the sea as soon as possible. The means of the control were continuous levee system, divergent canals and sediment control by Sabo works. In the 1950s, dams were added. All flood control works were built within rivers. Flood warning based on advanced meteorological observations and weather forecasts gradually became available through radio and, later in 1950s, on TV (Takeuchi and Shaw 2009b).

During the first 15 years after World War II, Japan experienced a series of devastating floods, which killed more than 1,000 people nearly every year. Such a tragic spell was caused by at least two reasons: wet climate and low maintenance of levees and channels during the war. The wet climate eventually ended and the flood control works gradually improved. It made the 1959 Isewan Typhoon, which incurred a death toll of 6,000 lives, the last flood that took more than a thousand deaths in Japanese modern history. It was obvious that the improvement in weather forecasting and dissemination by radio and TV greatly contributed to the decrease in casualties. Since the late 1950s, the floods became prevalent in the areas of rapidly urbanizing areas where paddy fields and hill slopes were converted to residential and industrial areas. This sprawling urbanization continued and floods in relatively small urban rivers and sporadic landslides became the major source of damages. The 1977 Nagasaki floods that took 375 lives caused by torrential rains of 180 mm/h (a record in hourly rainfall intensity in Japan) was an event attributed to extensive hillside development of urban areas.

In 1977, the River Commission of the Ministry of Construction recommended the "Comprehensive Flood Control Measures" in rapidly urbanizing basins. It suggested that not only retardation by dams upstream and river channel improvement downstream be undertaken, but also that the residential area should be developed so that no new extra discharge was generated. It promoted the installation of infiltration facilities and retardation facilities in newly developed residential areas as a measure. In 1987, the River Commission recommended the high standard levee policy and the construction of super levees started such as in the downstream of the Sumida River, Tokyo. A super levee is usually a land elevation behind the original narrow laser-like levee. The base length of a levee is typically 100–150 m long. The super levee protects the urban areas where extensive urban facilities stand and absolutely no levee breakage is allowed. The levee construction is part of urban redevelopment and the benefit is not only from the hinterland where the protection is provided but also the levee itself where the land value is increased due to redevelopment of the riparian areas.

The River Commission of the Ministry of Construction, Japan issued a new policy "Effective Flood Management, including Basin Responses" in December 2000 and passed the implementation role to the reorganized new Ministry of Land, Infrastructure and Transport. The new policy accepts flood inundation in habited areas through open dikes and secondary dykes. It is partly an extension of 1977 Comprehensive Flood Control Measures policy designated to rapidly developing areas in basins. But more substantially, it is a major policy shift from no floods in habited lands to some floods accepted in habited lands. The shift may be described as "from rivers to basin" as a strategy of confining floods within river lines by continuous levees. Dam reservoirs were changed to manage floods by river basin as a whole due to the fact that the recent increase in flood events and damages cannot be reduced only by the traditional non-inundation strategy.

In the present era, the construction of large dikes and dams with concrete is adopted as a major technological solution. This policy aims at no-flooding from the river. However, in earlier times, flooding from the river was considered a natural phenomenon in which people never tried to block the flood but rather developed some knowledge-based tools to reduce damages. The floods were considered also beneficial as they bring fresh coating of silt and mud helping soil regeneration and improved agricultural yield. Looking at the present threat posed by climate change, it is expected that mere technological options may not be sufficient to reduce disaster impacts. Moreover, some of these structures have failed in recent disasters and left people helpless. These experiences suggest that, in vulnerable areas, people should protect their life and assets by utilizing some historical indigenous knowledge and technology.

The above experiences reinforce the concept that the local governments alone cannot take care of the residents by alone. It is a participatory process, and the ownership of the process should remain with the local residents. The local information is important, and needs to be reflected in the community disaster prevention map and needs to be updated by the local residents themselves. The city and provincial governments need to play a facilitator role in this process.

Acknowledgements This chapter is based on different earlier work over more than 10 years, which is highly acknowledged.

References

Infrastructure Development Institute-Japan and Japan River Association (2006) Rivers in Japan. River Bureau, Ministry of Land, Infrastructure and Transport, Tokyo

Japan Sabo Association (2001) Sabo in Japan: creating safe and rich green communities. Sabo Publicity Center, Tokyo

Satoh T (2007) Study on evaluation of planning and responses to water-related disaster waste in Japan, and its application to Indian Ocean tsunami case in Sri Lanka. Master Thesis, Kyoto University, Kyoto

Suda Y (2012) Linking human behavior, participatory mapping and community infrastructure for disaster risk reduction: approaches in Shiso and Kamaishi. Master Thesis, Kyoto University, Kyoto

Takeuchi Y, Shaw R (2009a) Gender dimensions in risk communication: a perspective from a sediment disaster in Hiroshima, Japan, in Regional Development Dialogue. Gender Disaster 30(1):63–75 (Special Volume)

Takeuchi Y, Shaw R (2009b) Indigenous knowledge for river management and flood control in Japan. In: Shaw R, Sharma A, Takeuchi Y (eds) Indigenous knowledge and disaster risk reduction: from practice to policy. Nova Publisher, New York, pp 283–292

Uchida T, Nishimoto H, Osanai N, Shimizu T (2009) Countermeasures for sediment-related disasters in Japan using Hazard Maps. Int J Erosion Control Eng 2(2):46–53

Umakoshi H (2011) Level of resident participation and the process of implementing the "Act on Sediment Disaster Countermeasures" in Asaminami, Hiroshima City. Master Thesis, Kyoto University, Kyoto

Yoshida Y, Takeuchi Y, Shaw R (2009) Town watching as the useful tool of urban risk reduction: case of Saijo, Japan. In: Shaw R, Srinivas H, Sharma A (eds) Urban risk: an Asian perspective. Emerald, Bingley, pp 189–206

Part II
Community Based Risk Reduction Issues

Chapter 4
Social Capital in Disaster Recovery in Japan: An Overview

Nozomi Hishida and Rajib Shaw

Abstract Generally, Social Capital refers to networking, trust, and norms among people, which can be formed in the society. It is supposed that being rich in social capital, in other words, having strong faith among citizens (or members which configure a community) and having strong ties inside communities positive effects, such as enhancing political commitment, improving performance of government and businesses to be more effective, improving educational performance, public security and stimulating the local economy. Although, social capital can give negative impacts to recovery processes sometimes, generally the strong state in social capital can contribute to the recovery of communities affected by disasters. That's because social capital facilitates various activities and enhances performance of social actors, which are essential for disaster recovery.

Keywords Community recovery • Disaster prevention • Disaster recovery • Social Capital

4.1 Introduction

On March 11th 2011, Tohoku region (North-East part of Japan) was hit by a cataclysmic catastrophe caused by tremendous earthquake and tsunami. It washed out numbers of towns, and it took thousands of lives. The damaged area was too broad for governments' help to reach to each town in short duration. The local governments of devastated area themselves became a victim and every function in Pacific coast Tohoku area was lost. The worst of it was the crisis at Fukushima Nuclear Power Plant No.1, which brought disruption to central government.

N. Hishida (✉) • R. Shaw
Graduate School of Global Environmental Studies, Kyoto University, Kyoto, Japan
e-mail: hishida-nz@n-koei.jp; shaw.rajib.5u@kyoto-u.ac.jp

R. Shaw (ed.), *Community Practices for Disaster Risk Reduction in Japan*, Disaster Risk Reduction, DOI 10.1007/978-4-431-54246-9_4, © Springer Japan 2014

Thus, at the very initial period (assumed as several days), the devastated people hardly got the governments' help. Under such conditions without governments prompt response, the victims had shown great patience and undergone unbearable fact by helping and supporting each other. Foreign mass media featured the people who formed a line in front of a grocery shop without making a fuss and looting and the world gave plentiful applause to these calm, mannerly victims. Why they could believe, help and rely on each other in the extreme condition?

The concept of social capital refers to trust, norms, tie and networking among people in a society. After the publication of Robert D. Putnam's notable work "Making democracy work" in 1993, it has been often discussed in context of policy studies, economics, business administration studies and sociology as it can bring certain impact on performance of the society (it can be effectiveness of governance, economic, business administrative and about other social issues).

On the other hand, in terms of disaster management, there is insufficient number of studies which directly discuss on social capital in the context of disaster preparedness and recovery. However, after Great Hanshin earthquake, there are several studies which discuss about volunteer's contribution and community disaster management. Moreover, Great East Japan earthquake remind us the importance of "Kizuna" (tie) among people to overcome the disaster, which is absolutely correspond to the importance of social capital for disaster management. Hence, by reviewing the past disaster experiences from social capital aspect, we can learn more about "Kizuna" in a community and a society. In this chapter, we will focus on three types of social capital; bonding social capital among neighbors, bridging social capital between in and out of damaged area, and trust between residents and government. The key words are "disaster reduction", "recovery of damaged communities" and "support from external community".

4.2 Overview of the Concept of Social Capital

4.2.1 General Definition of Social Capital

Generally, social capital refers to networking, trust, and norms among people, which can be formed in the society. There are various definitions of social capital and its components and functions. Compared to natural capital, economic capital, and human capital, social capital is vastly difficult to be calculated and evaluated. One definition of social capital provided by Social Capital Initiative, World Bank is "*Social capital refers to the internal social and cultural coherence of society, the norms and values that govern interactions among people and the institutions in which they are embedded. Social capital is the glue that holds societies together and without which there can be no economic growth or human well-being.*"

It is said that the concept of social capital was highlighted the first time in a study of Bourdieu. He defined the concept as "*the aggregate of the actual or potential resources which are linked to possession of a durable network of more or less institutionalized relationships of mutual acquaintance or recognition*" (Bourdieu 1985). He focuses on the benefit delivered to individuals by virtue of participation in groups and on the construction of sociability for the purpose of creating these resources. The amount of social capital resources people have depends on the scale of the network which each individual join and on the amount of economic, cultural, and social capital owned by the members of the network. The network here refers to families and relatives, personal network and connection, etc.

Coleman, on the other hand, gives rather vague definition. He pointed out that social capital is defined by its function. "*It is not a single entities but a variety of different entities, with two elements in common; they all consist of some aspect of social structures, and they facilitate certain actions of actors— whether persons or corporate actors — within the structure*" (Coleman 1988). For Coleman, social capital can encompass not only family and relatives but also local network as community and norms which is necessary to form and keep the community. In his theory, Coleman mentions two aspects of social capital; one is the structural aspect of social capital, which is indicated in the definition above, and another is cognitive "factors" which exist in the society. Coleman specially details following factors as social capital:

- Obligation, expectation, and trustworthiness of structure
- Information channel
- Norms and effective sanction

This view of social capital captures not only social structures at large but the ensemble of norms governing people's interpersonal behavior (Serageldin and Grootaert 2000).

After Coleman, Putnam presents the most famous definition of social capital. According to Putnam, "*social capital refers to features of social organization, such as networks, norms, and trust, that facilitate coordination and cooperation for mutual benefit*" (Putnam 1993). In his famous work in Northern and Southern Italy, Putnam (1993) describes the society which has rich social capital:

> Some regions of Italy, such as Emilia-Romana and Tuscany, have many active community organizations. Citizens in these regions are engaged by public issues, not by patronage. They trust one another to act fairly and obey the law. Leaders in these communities are relatively honest and committed to equality. Social and political networks are organized horizontally, not hierarchically. These "civic communities" value solidarity, civic participation, and integrity. And here democracy works.

Norris (2002) states that most importantly Putnam's theory clearly mentions two phenomena of social capital: one is the structural phenomenon (social networks) and the other is the cultural phenomenon (social norms). Meanwhile, Putnam's definition is often said as the narrow, specific definition on the point that his definition focuses on horizontal network among people, which give positive impact to society and democracy. Serageldin and Grootaert (2000) stated that the definition presented by Coleman cover broader area. They mainly point out that Coleman's concept includes vertical association as well as horizontal, and this wider range of association covers both positive and negative objectives. Fukuyama (1995, 2001) discusses social capital under an even broader context. He simply defines social capital as "an instantiated informal norm that promotes co-operation between two or more individuals" (Fukuyama 2001). He shed light on trust and says that social capital can give positive impact on economy in a society which has strong social trust.

4.2.2 Social Capital and Governance

Social capital is often discussed because of its relevance to governance. Regarding social capital and governance, Putnam's work in Northern and Southern Italy is possibly the most famous. Putnam's explanation for the difference between Northern and Southern Italy on their governance performance is concluded as the difference come from the type of social capital they have. Putnam mentions that the bridging type of social capital is rich in Northern Italy and it helps governance works well. This study is very significant on the point that it clearly describes the relation between social capital and governance in the case study. However, his description may need more analysis on interaction between social capital in the civil societies and the governmental sectors. His theory states about the impact, which is indirectly given from the civil societies to the governmental sectors but not about the effect which governmental sectors give to the civil societies. There is a lack of viewpoint of vertical interaction between civil societies and governmental sectors.

In their study on social capital and urban governance, Maloney et al. (2000) discuss about the importance to consider top-down interaction perspective, interpenetration of public authorities and voluntary community associations. It is said that research on social capital should not focus on only the effect of community-level social capital on government performance, but also the effect of governmental associational relationships on social capital. Here, the need to supply the top-down perspective to more common bottom-up, horizontal perspective suggested by Putnam. Regarding how social capital affects the relationship between government sector and civil societies, Evans (1996) propounds the "state-society synergy." By this concept, he argues that active government and mobilized communities can enhance each other's developmental efforts. What his study claims is that norms and cooperation and networks of civil engagement among citizens, which are core of social capital, can be promoted by public agencies and used for developmental ends.

Creative actions by government organizations can foster social capital; linking mobilized citizens to public agencies can enhance the efficiency of government. The combination of strong public institutions and organized communities is a powerful tool for development. Better understanding of the nature of synergistic relations between states and society and the conditions under which such relations can most easily be constructed should become a component of future theories of development (Evans 1996). Importantly, Evans mentioned that synergy is constructible, regardless the condition of the pre-existed social capital, even in the more adverse circumstances typically found in developing countries.

On the contrary, Bowls and Gintis (2002) take a stance of that government and civil society relationship can be explained as complementally relationship.

He states that well-working communities require a legal and governmental environment favorable to their functioning, and also says that governmental intervention has sometimes destroyed community governance capacities. So if government and community have well-understanding for each other, community governance can be easily enhanced rather than in case there is lack of communication between them. Meantime, government intervention cannot work desirably without understanding of local context, but tend to give damage to preexisted community governance.

4.2.3 Bonding-Social Capital and Bridging-Social Capital

Above all, the most famous and important perspectives for social capital are "bonding social capital" and "bridging social capital". Bonding refers to the value assigned to social networks between homogeneous groups of people such as territorial groups, neighborhoods, disciplined organizations etc., and Bridging refers to that of social networks between socially heterogeneous groups which are more open to other member of a society who have different interests, ideas, approaches and activities from them. Bridging social capital is argued to bring benefits for societies, governments, individuals, and communities. While bridging social capital which connects different actors in a society is often described to contribute for social utility, bonding social capital tend to have exclusiveness and it is likely to be stated in negative context such as criminal gang etc.

The understanding that these two types of social capital give different influences to a society and its members is shared among many persons. However, when it comes to disaster management, bonding social capital, which enhances community ties is supposed to bring positive result. This point will be mentioned in next section.

4.3 Pre-disaster Preparedness and Social Capital

In his study on the role of social capital in disaster risk management, Ueda (2011) noted the five perspectives to deal with disaster risk.

a) Minimizing human damages in immediate after disaster
b) Recovery from proprietary loss by disaster
c) Recovery from emotional and psychological loss
d) Risk information sharing in times of peace
e) Government's measures for disaster vulnerability of lands and cities

Among those, the perspectives which deeply relate to disaster prevention, risk reduction and social capital is: a) Minimizing human damages in immediate after disaster and d) Risk information sharing in times of peace [a) is actually explains about people's action at the last minutes and immediate after the disaster, but it can be categorized as prevention and reduction in terms of minimizing possible damages and losses].

4.3.1 Minimizing Human Damages Immediately After a Disaster

Minimizing human damages in immediate after a disaster means how many lives will be saved after a disaster happen.

On the occasion of devastating disaster such as the Great Hanshin Earthquake and the Great East Japan earthquake, it is difficult for public rescue teams to cover all of damaged area in brief period. At that time, it will become the difference between life and death whether neighbors, the disaster victims themselves can help each other or not. In such case, the people who possibly help you and the people who possibly you can help, they are likely to be the one who are in your neighborhood or at the same school or workplace with you when a disaster happen. It supposed that bonding social capital among the members in neighborhood, same school and same organizations call up people's resolve to rescue others.

4.3.2 Risk Information Sharing in Times of Peace

Risk information sharing basically means risk communication. Trust between information provider and receiver is crucial for effectiveness of risk communication. (Who believe the information from unreliable person?) Anybody can be an information provider as long as the information is accurate. Though, in Japan, especially in terms of local disaster risk information, the governments (National and local) tend to take the main role. So, it is required that there is not only horizontal trust among residents but also mature sense of vertical trust between residents and government. It should be noted that trust between residents and government is differ from the resident's dependence on government.

There have been many notes about heavy dependence of residents on government as result of attitude of people who are obsessed with the idea that government would do everything for them in the emergency. However, as the case of the Great Hanshin Earthquake clearly shows, when a disaster suddenly happens and you are exposed to risk, it is not always come that government's team would be near your place to save your life.

What makes the trust between residents and governments possible is government's proper supports and resident's independent attitude what comes with their own appropriate judgments and actions. When disaster strikes, the first responders are not trained emergency personnel but rather local residents and neighbors (Perrow 2007), and at the policy level, it is required to recognize the social capital of the communities as an asset (Nakagawa and Shaw 2004).

4.3.3 Case of Nanmoku Village

The good example of the minimizing of human damage and risk information for disaster response is the case of Nanmoku Village. Nanmoku village is a small, highly aging village in a mountainous area in Gunma Prefecture, Kanto region, Japan.

The typhoon in September 2007 which hit Gunma Prefecture and accompanied sediment disasters made this small village isolated, and at least temporary, they were in condition which they could not expect external support. There could be considerable damage under the situation, but in fact, it ended up with zero victims in the village.

It is considered that what made it possible to minimize the damage were appropriate judgments by village office and active responses by residents. "When a sediment disaster alert was given, the officers of village office judged that it would be more dangerous for elder villagers to evacuate long way from their places to village office under the heavy storm. They made an announcement over the community wireless system again and again to tell villagers that village office saffs would soon to be not able to go to their place to help them and asked them to help and support to each other to get through the danger, instead of give them a call for evacuation. Upon the village office's request, the villagers help took actions: to sandbag for single elderly people, to pick up the people who live near the river by small trucks, etc."

Giving a call for evacuation is an ordinary response for public administrations, in case of given a sediment disaster alert. However, villagers in Nanmoku village was elders and it was very hard for them to evacuate to the official shelter where is far from their houses. The village was so small and many of them had lived there for long time and officers had close contact with their people. The officer knew about their village and people well and it was clear that some people would be in danger on the way to shelters from their houses (Oikawa et al, 2008, Katada, 2009). Their environment with strong territorial bonding and decades of neighborly ties was likely to foster bonding social capital rather than bridging social capital.

As stated before, bonding social capital tends to be stated in negative contexts. But in this Nanmoku village case, bonding social capital among villagers included village officer enhanced mutual faiths and supports. The disaster response in Nanmoku village would have not work without people's mutual support and trust between residents and local government (village office).

At this point, it can be noted that bonding horizontal capital among residents and vertical social capital between residents and local government (village office) take important role in disaster management especially in disaster prevention and risk and damage reduction.

4.4 Post Disaster Recovery

4.4.1 Former Discussion on Social Capital and Post Disaster Recovery

In terms of disaster recovery, Aldrich (2010) mentioned that Social capital—the networks and social resources available to us—matters immediately following crisis (the emergency response phase) and in the long period of recovery afterwards (restoration). Nakagawa and Shaw (2004) pointed that individual efforts is essential for the satisfied rehabilitation, and disaster recovery which means not only building houses but the reconstruction of the whole community as a safer place needs to mobilize each member of the community in collective action. Based on their case studies in Kobe and Gujarat, they found that social capital is one of the most effective, crucial elements in enhancing collective action and disaster recovery.

After the experience of Great East Japan Earthquake, Yamauchi (2011) discussed about local groups for disaster reduction activities (Jisyu-Bousai-Soshiki), volunteers and NPOs in context of social capital types. Jisyu-Bosai-Sosiki is based on local communities and it related to bonding social capital. On the other hand, the people who join NPO works connected each other by common purposes and interests, which contributes to foster bridging social capital. Needless to say, both types of groups and activities take important roles for disaster reduction and recovery, and Yamauchi mentioned about the possibility of new type of social capital which is integration of bonding and bridging social capital.

Inaba (2011) also stated the social capital in terms of trusts and cooperation, which disaster victims of Great East Japan Earthquake showed after the disaster and claimed that social capital supported the victims during the hard time after the disaster while the government could not implicate adequate policies and measures. He discussed on implication of policy to develop social capital, and mentioned that Education, especially public schools as community centers, gives affect on social capital in communities. Besides, he pointed that it is important work for the government to give the opportunities to local (internal) communities to contact to external communities to prevent negative impacts of social capital. Sakamoto and Yamori (2009) indicated that in some major factors relating life recovery, on a long-term basis, the most important factor was characterized by the term "social ties," specifically referring to human networks.

As all authors above mentioned, one of the major role of social capital for disaster recovery is to support the disaster victims under difficult situation and to contribute to the recovery of their community by connecting the community members. As Yamauchi mentioned, this function is generated from the bonding among people in a community. The point of view has been noted in terms of community disaster management in various past disaster experiences in Japan. Actually, the concept of social capital has often been discussed in context of community disaster management, and in that case, it has often meant bonding social capital.

Moreover, the experience of the Great East Japan Earthquake generated a new perspective. That is to consider the supports from volunteers and NPOs/NGOs as bridging social capital. In disaster recovery process, the hope and actions of suffered people for rebuilding their own community are essential engines toward post disaster recovery. Also, bonding social capital takes important role in the period of recovery; the period that more gaps arise among victims, who have different degrees of damages, official supports etc. Under this complicated and sensitive situation, bonding social capital which mobilize the community member and enhance their collective actions should be considered as a big factor for post disaster recovery.

On another front, in case of catastrophe accompany with devastating and extensive damages, the damages of human loss, economical loss and infrastructure loss are so large that the devastated areas cannot archive prompt recovery without supports from external resources. Wherein it is desired that the function of bridging social capita which create the relationships with volunteers, NPOs/NGOs and other kinds of external resources, and encourage the suffered people to receive their supports. Following sections will describe the role of these two types of social capital in post disaster recovery process.

4.4.2 *Function of Bonding Social Capital as Community Ties*

In this section, we would like to introduce the case of temporary housing of the Great Hanshin Earthquake as the example of the function of social capital. After the Great Hanshin Earthquake, about 32,000 temporary housings were provided in Kobe city. Many of them were built not in devastated Kobe city but in suburb areas. With the government's policy which set great store on quick transfer of the elders and the disabled from evacuate shelters to temporary housing, they were given priority transfers. As the result, they had to settle in temporary housing torn apart from the linkage with their own community and neighbors. It meant that the people who had weaker living basics became the first to lose the local community which had supported their lives, and it ended up to numbers of solitary deaths in isolated temporary housings (Kobe City, 2012).

This tragedy opened up to the movement to take different approaches of temporary housing, trying to keep the existing communities as much as possible, or when it was difficult, create opportunities which could alternate the functions of original communities, such as neighborhood gatherings and institute systems for temporary housing management. This leaning from the Great Hanshin Earthquake has been applied to other disasters such as Chuetsu Earthquake in 2004 and Chietsu-Oki Earthquake in 2006, and the Great East Japan Earthquake.

As noted previously, bonding social capital which is generated in neighborhood communication brings peoples' mutual helps, supports victims and enhances actions for recovery. And it is one of the few resources which suffered people who lost economic and human worthiness from disaster could rely on. Losing the linkage, it can be described as secondary disaster for them.

At this point, the people who currently suffered from the most is the people near Fukushima No. 1 Nuclear Plant. Their towns have been specified as restricted areas, and they have been forced to leave their houses and evacuated to other cities and prefecture separately. Supporting quick transferring of the victims from evacuate shelters to temporary housing and maintaining original communities at the same time is quite difficult task for local governments, indeed. But we should always remember that bonding social capital in own community is invaluable "capital" for victims.

4.4.3 *Function of Bridging Social Capital as Connection to External Resources*

Meanwhile, in terms of long-span recovery of broader area like towns, cities and could be regions, it seems that the more damages they have got, the more bridging social's role become important. In the Great East Japan Earthquake, local governments themselves were disaster victims. It is almost impossible for them to solve all the problems and archive all tasks on years of long way for recovery by themselves with limited resources in divested area. Hence, it has a crucial meaning for them to take in supports from volunteers, NPOs/NGOs, other municipality and many other external resources, for years of long process of post disaster recovery.

In this regard, it supposed that bonding social capital in a community does not function in a positive way, because bonding social capital often associates with exclusivity. So, in a different way from ties in neighborhood, to have bridge with people who are physically distant would be a safety net in emergencies. It remains a lot of matte of discussion on how this bridging social capital is fostered, in what kind of forms and activities. It may be possible for individuals to have personal connection with private groups for sharing same interests. It may be possible for local communities to have connection to NPOs etc. in times of peace. Every topics and interests can be a seed for a bridge between them.

Apparently, it is difficult to promote the actions of individuals and each local community according to a plan. But, in this respect, with more observation and discussion, it can be indicated that it will work effectually in emergencies for municipalities to have relations with external resources and foster bridging social capital.

4.5 Future Perspective

Disaster management requires wide range of approaches in every level of society, from a measure by national government to neighborhood activities. As it was shown in previous sections, bonding social capital in one community and bridging social capital which creates the connection between a community and external groups are taking a notable role both in disaster reduction and post disaster recovery. In order to cover a society's disaster vulnerability, to cultivate their resilience, measures and activities, which foster both types of social capitals need to be boldly promoted. Figure 4.1 shows how social capital plays an important role in disaster management.

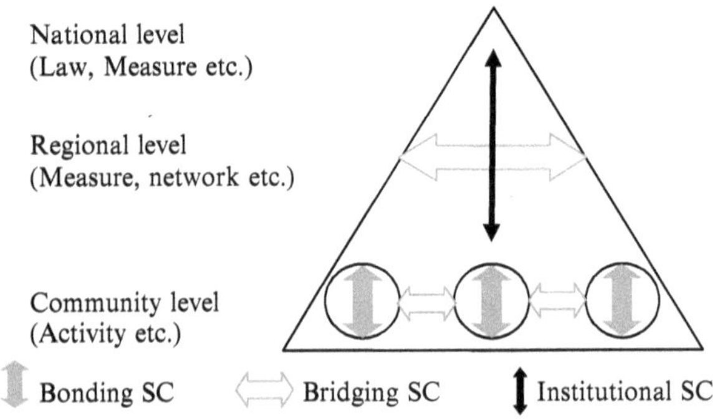

Fig. 4.1 Social capital approach in disaster management

Two typical forms of bonding social capital are neighborhood community association and local volunteer groups for disaster prevention (Jisyubosai Soshiki). Generally these activities are more active in rural area where they have more solid, traditional community management system. Today, in terms of this community management system, there are problems both in rural and urban area. In rural area, numbers of communities are in danger of disappearing because of depopulation, while many communities are often to be quite unable to function in urban area due to lack of neighborhood communications. The approaches and goals for a good community may be different in rural and urban area as people's relations and lifestyles are greatly different. But it is clear that, for the aspect of disaster management, we need the function which develops bonding social capital to save ourselves, and local government and residents should do some actions to keep the function of communities. It can be grassroots activities and it is fundamental and essential for a safe society.

Meanwhile, the approaches to develop bridging social capital between a community and outsides can be promoted by local autonomy in a strategic way of making real connections with external autonomies, NGOs/NPOs and other people. Needless to say, individuals and groups can have connection to outsides and such opened atmosphere is exactly the soil of bridging social capital. Yet, not only the mood, but it is ideal to have concrete figures of "bridge". The topics of these bridges are not necessary always to be disaster; they can be about environment, tourism, leisure, foods, volunteer works, and so on. The focal point is that they have good connection and relations with external groups. With the ideas of bonding and bridging social capital, the well-prepared and resilient society will be cohesive and open society. It sounds paradoxical, but these different kinds of a society will support itself from inside and outside.

Acknowledgements The authors acknowledge the financial support of CWS and MERCY Malaysia to conduct fieldwork in Tohoku disaster affected areas.

References

Aldrich D (2010) Fixing recovery: social capital in post-crisis resilience. J Homel Secur 6:1–10
Bourdieu P (1985) The social space and the genesis of groups. Theory Soc 14(6):723–744
Bowls S, Gintis H (2002) Social capital and community governance. Econ J 112:419–436
Coleman J (1988) Social capital in the creation of human capital. Am J Sociol 94:S95–S120
Evans P (1996) Government action, social capital and development: reviewing the evidence of synergy. World Dev 24(6):1119–1132
Fukuyama F (1995) Trust: the social virtues and the creation of prosperity. Free Press, New York
Fukuyama F (2001) Social capital, civil society and development. Third World Q 22(1):7–20
Inaba Y (2011) How to develop social capital – a perspective from policy making (Shakai Kankei Shihon wo Dou Jousei Surunoka – Seusaku Taisyo to Shiteno Shiten). DIO Research Institute for Advancement of Living Standards Report (DIO RengoSouken Report)
Katada T (2009) Keynote speech in disaster symposium in Mukogawa, 21 Feb 2009
Kobe City (2012) Overview and recovery: Hanshin-Awaji Great Earthquake (Hanshin-Awaji Daishinsai no Gaiyou Oyobi Fukkou). Kobe City (Kobe Shi.) http://www.city.kobe.lg.jp/safety/hanshinawaji/revival/promote/

Maloney W, Smith G, Stoker G (2000) Social capital and urban governance: adding a more contextualized 'Top-down' perspective. Polit Stud 48(4):802–820

Nakagawa Y, Shaw R (2004) Social capital: a missing link to disaster recovery. Int J Mass Emerg Disasters 22(1):5–34

Norris P (2002) Democratic phoenix: reinventing political activism. Cambridge University Press, Cambridge

Oikawa Y et al (2008) Importance of community-based response to storm and flood. In: Conference Report of Japan Society of Civil Engineers ("Chiiki Community ni Oujita Gouu Saigai Taiou no Jyuyosei – Heisei19nen Taifu19gou Saigai ni Okeru Genkai Syuraku to Seijuku Comyunity ni Okeru Jirei", Dobokukeikakugaku Kenkyu, Koen Syu)

Perrow C (2007) The next catastrophe. Princeton University Press, Princeton

Putnam R (1993) Making democracy work: civic traditions in modern Italy. Princeton University Press, Princeton

Sakamoto M, Yamori K (2009) A study of life recovery and social capital regarding disaster victims – a case study of Indian Ocean Tsunami and Central Java earthquake recovery. J Nat Disaster Sci 31:49–56

Serageldin I, Grootaert C (2000) Defining social capital: an integral view. In: Picciotto R, Wiesner E (eds) Evaluation & development: the institutional dimension. Transaction Publishers, New Brunswick

Ueda K (2011) Soft control in disaster risk management – the role of social capital (Saigai Risk Management ni okeru Soft Controle, Social Capital no Yakuwari). J Soc Capital Res (Shakai Kankei Shihon Ronbunsyu) 2:29–48

Yamauchi N (2011) Role of social capital in disaster prevention and recovery (Bousai, Saigai Fukko ni okeru Social Capital no Yakuwari). DIO Research Institute for Advancement of Living Standards Report (DIO RengoSouken Report)

Chapter 5
Concepts and Approaches of School Centered Disaster Resilient Communities

Shohei Matsuura and Rajib Shaw

Abstract Mega disasters, such as the East Japan Earthquake and Tsunami (EJET), have devastating effects on schools and communities. Not only do they interrupt educational activities, but also weakens community ties due to displacement of affected people that exacerbates the chronic low birth—aging problem faced by many rural communities in Japan. This paper intends to rationalize the concept of School Centered Community Building that is proposed by the Ministry of Education, Culture, Sports, Science and Technology (MEXT) of Japan as a possibility to restore the community ties weakened by EJET though schools, by first looking into new ways to utilize the schools by making them multi-functional facility aimed to strengthen school–community linkage. It will then examine the approaches of community participation in school management and educational activities to realize school-centered community building. Finally, the paper will analyze the benefits and challenges of this concept to come up with possible ways to advance the school recovery process.

Keywords Community building • Disasters • East Japan Earthquake and Tsunami • School centered recovery

5.1 Introduction

The East Japan Earthquake and Tsunami (EJET), which occurred at 2:46 pm on 11 March 2011 with magnitude of 9M, left 15,867 dead and 2,904 missing (Ministry of Education, Culture, Sports, Science, and Technology (MEXT) situation report, as of 25 July 2012). EJET Japan alone cost US$210.0 billion or 57.4 % of global reported

S. Matsuura (✉) • R. Shaw
Graduate School of Global Environmental Studies, Kyoto University, Sakyo-ku, Yoshida-Honmachi, 606-8501 Kyoto, Japan
e-mail: matsuura.shohei.8s@kyoto-u.ac.jp; shaw.rajib.5u@kyoto-u.ac.jp

R. Shaw (ed.), *Community Practices for Disaster Risk Reduction in Japan*, Disaster Risk Reduction, DOI 10.1007/978-4-431-54246-9_5,

damages for the whole year, marking it as the most expensive natural disaster in terms of economic damages in 2011 (Guha-Sapir et al. 2012). Among the five prefectures in the East Japan or Tohoku Region (Aomori, Iwate, Miyagi, Fukushima, Yamagata and Akita), the most heavily affected prefectures were Iwate, Miyagi and Fukushima, in which Iwate and northern area of Miyagi share a Ria coastline, often characterized as having a zigzag coastal formation and southern area of Miyagi and Fukushima have a plain land. The different geographical features resulted in contrasting effects of tsunami damage. Significant numbers of survivors are still displaced to unfamiliar locations inside and outside of their former residential cities and regions. As of July 2012, the number of displaced people totaled 344,000 in which 71,510 are currently living outside of their prefecture in evacuation facilities, hotels or relative's homes (Reconstruction Agency, Japan 2012).

As for damages in the education sector of these three prefectures, 653 deaths, 91 injuries and 75 missing of students and teachers have been reported (number includes both public and private kindergarten, elementary, junior high and high schools, special needs school, vocational schools and universities) and over 2,400 education related facilities are still in need for rehabilitation in different levels. Twenty-five thousand five hundred and sixteen students have been transferred to another school, while the three Tohoku prefectures have over 14,000 students who transferred schools outside of their former residential prefecture (MEXT 2012a, b). Majority of the students who transferred outside of their prefectures are most likely from Fukushima where they had to evacuate outside of the 20 km radius exclusion zone from the effects of radiation. Educational activity, on the other hand, has resumed fairly quickly, currently close to 100 % for public schools. These include education for students who have transferred out of their original schools inside and outside of the Tohoku Region.

When faced with such mega disasters, communities play a very important role in responding to them, particularly at the local levels. Local communities, which usually become the first responders, must make response decisions by themselves with family members, relatives and neighbors, as external assistance may not be available (Haghebaert 2007). Even with good risk reduction planning, it is ultimately the community ties, trust and coordination built during normal times that greatly enhance the self-help and mutual-help aspects of community based disaster risk reduction. In this sense, school plays a prominent role in building this community tie because schools commonly functions as a cultural and spiritual center for the community for different age groups, together with being a facility to provide education for children (Takeuchi et al. 2011). For this reason, implication of EJET's effect on the education sector has been immense not only for school children and teachers, but on the whole society.

In addition, many schools are designated as evacuation centers, not only for their safe location, but because many community members trust school as a stronghold to flee from danger and where they can meet up with their family and neighbors during emergencies, after evacuation. Disaster risk reduction (DRR) education and other activities conducted in schools also connects schools with communities because DRR education is not confined in schools, but involves different aspects where the whole community must take collective actions. In this respect, Shaw et al. (2011) state that the key to disaster education is the linkage between school and community, community and family, different disciplines, stakeholders, nature and human. Hence, effects on schools from EJET have paralyzed the feeling of safety and security in the community.

The consequence of EJET on population decline due to displacement has been mentioned above. However, the existing problem of low birthrate, aging and migration that is more noticeable in the rural regions of Japan, is even a larger bottleneck that would threaten early recovery and community building in the Tohoku Region. Although the issue of population decline was already recognized as a social problem in Japan since the 1990s, the effects of EJET have aggravated this problem. For schools, population decline is an alerting issue that could close down schools, as there will not be enough students to operate the schools. For the community, Ishiwatari (2012) mentions on an example that the declining population of younger age group has been reducing the number and capacity of voluntary community based disaster management groups, such as Syobo-dan (firefighting), Suibo-dan (flood fighting) and Jisyubo (earthquake disaster management), weakening disaster resilience of communities, including schools. This in conjunction with disruption of school's education programs, particularly of those that involve community participation is another bottleneck that must be taken measure to maintain and strengthen community ties.

Kyodo News Agency on 22 June (2012) reported on the MEXT White Paper 2011 affirming that rehabilitation of school, a central public facility of the community, is the key in revitalizing community ties. Based on this argument, MEXT presented its policy on "School Centered Community Building," that suggests in making school a multi-functional facility, such as by combining community center, library or warehouse for emergency goods in which the whole community can jointly utilize, the new school would become a hub for community interaction as well as for disaster risk reduction that may well encourage residents to stay or for dispersed people to return to their communities. However, even though the concept is much encouraging for the recovery of the school and community, implementation of this policy is still minimal on the field.

The reason for this may be explained with the fact that the discussions still focus on constructing the school building, which is relatively a short-term goal of school recovery. While it is understandable that the structural recovery of schools must be prioritized and requires substantial time and budget, discussions on the roles of schools for community building, which is a much longer process that could take up to 5–10 years, must be simultaneously considered, so that the community will be able to fully utilize the new school facility to rebuild community ties. As of this stage, the responsibility for school recovery largely belongs to the Department of Facilities Planning and Administration of MEXT, which is responsible for the structural aspect of public schools. However, identifying the roles that school will take in the overall recovery process and community building will require other ministries and departments, both at central and local levels, to work together. This is because when considering the non-structural aspects and long-term goals in school recovery, policy makers and implementers will be faced with such interdisciplinary issues as integrating hard and soft measures in DRR, methods to strengthen community ties through school–community linkage through education and school management reforms.

With this in view, this chapter will attempts to rationalize how the "Building School Centered Disaster Resilient Community" concept will be an effective way to rebuild and strengthen community ties by providing possible measures to overcome the issue of declining population caused by displacement from the effect of EJET and low birth—super aging society and interruption of educational activities. The chapter first sheds light on the community profile of the Tohoku Region, specifically on the population decline issue caused by displacement, low birthrate—super aging and migration that impose significant implications on school recovery. Following that, the chapter looks into the concept of "Community School" to observe the different approaches in school management and education with community participation. Finally, the chapter provides the details of the "School Centered Community Building" concept of MEXT that can be adopted to overcome the bottlenecks in building a school centered disaster resilient communities in the Tohoku Region. Successful and sustainable implementation of this concept may become the key to push forward the overall recovery process in the Tohoku Region.

5.2 Population Decline and Its Implication on the Education System and Communities

5.2.1 Background

The importance of examining the social profile of the Tohoku Region lies in the fact that school recovery will be affected by the demographic changes, especially concerning the transformation of ratio of different age groups. As mentioned in the introduction, the continuous population decline in Japan has been a serious concern that threatens community ties, thus affecting DRR capacity to weaken. For example, around 65 % of the fatalities of the EJET were people over 60 years old (Cabinet Office and Government of Japan 2011), which have brought concerns for the aging communities, because evacuation during emergencies and medical care in the recovery phase for the elderly population, who are physically weak, becomes a major challenge (Shaw and Takeuchi 2012).

As already mentioned, the Tohoku Region is currently facing significant change in its demographic profile because of displaced people affected by EJET. The effect of EJET escalated population decline in the three most affected prefectures of the Tohoku Region. Figure 5.1 displays the overall trend of the population change by showing the number of population moving out subtracted by the number of population who moved in. This figure does not include others who have moved within their prefecture most noticeably from the affected coastal areas to the inland area. In detail, Miyagi Prefecture shows major population decline in the coastal area, such as Ishimaki (−6.55 % or −10,508 people during March 2011–2012), but Sendai (+0.68 % or +7,076 people) and new town area such as Rifu, located inland of the

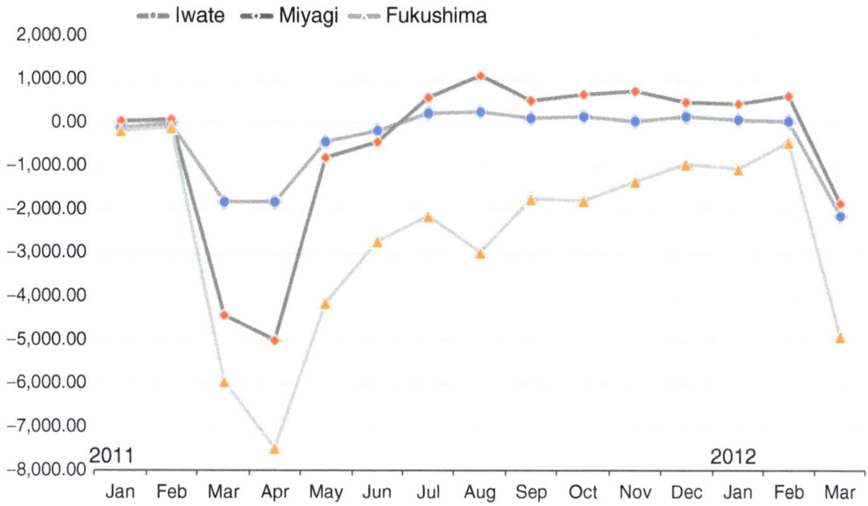

Fig. 5.1 Number of population moving out minus moving in (Iwate, Miyagi and Fukushima). Adopted from data and information from the Ministry of Land, Infrastructure, Transport and Tourism White Paper (2011)

prefecture, has experienced population increase (Tohoku Bureau of Economy and Industry 2012). These migrations from and to specific areas should also be grasped to see the whole picture. Although it is hopeful that these displaced people will return to their respective communities, if the recovery process becomes overly extensive, these population movement could permanently change the demographic profile of the region.

On the long run, demographic changes with fewer children due to low birthrate and increase of age group over 65 years of age, known in Japan as *shoshika* and *koreika* respectively, will gradually show a deeper impact to the social profile. Another factor is the migration of working age group (age over 15 and under 65) moving out from the non-urban areas to urban cities with their children to seek job and educational opportunities, leaving the elderly people in small cities and townships. Regarding the population drainage from non-urban to urban areas, the National Statistic Report in 2010 shows population in 9 prefectures (mostly urban areas, including Tokyo and Osaka) has had population increase while 38 prefectures and cities lost their population. This polarizing trend in population is another major factor that is contributing to the increase of fewer children—aging community problem in the non-urban regions. Although the population decline problem had existed before EJET, the after effect of the disaster is expected to further accelerate this situation. The combination of these three factors, *shoshika*, *koreika* and migration, occurring concurrently is and will be the cause of continuous population decline and a major concern for losing regional social and economic vitality.

To briefly provide an overview of the *shoshika* problem in the Tohoku Region, statistics show that since the population peaked in the region at 9.83 million in 1995, it has steadily declined to 9.34 million in 2010. The National Institute of Population

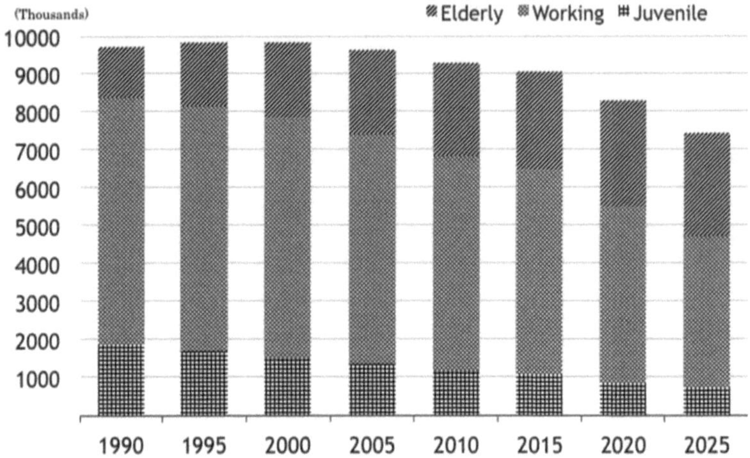

Fig. 5.2 Transition of age groups in the Tohoku Region (1990–2025). Created from Statistics Bureau of Japan (2101) and National Institute of Population and Social Security Research (2011)

and Social Security Research projects that the population will further decrease to 9.06 million in 2015, 8.29 million in 2025 and 7.43 million in 2035. As for birthrate in the post-war period, the average in the Tohoku Region peaked at 4.38 (295,940 births) in the 1950s and since then continuously declining to 2.07 in the 1970s, 1.66 in the 1990s and 1.41 (70,047 newborns) in 2010, meaning that the number of newborns in the Tohoku Region is less than 1/4 compared with the peak in the 1950s. As for the percentage of age groups, which is divided in three groups, juvenile age group (under 15), working age group (from 15 to 65) and elderly age group (age 65 an above). The ratio of these age groups and the characteristic of the demographic changes of the Tohoku Region in the recent years can been seen in Fig. 5.2. Students of elementary and junior high schools are represented in the age group under 15.

5.2.2 Shoshika and Transformation of School Facilities

Shoshika, as described above, poses a serious threat to schools to close down or be merged with other schools, especially of those in smaller cities in non-urban regions, because there will not be enough students for schools to operate. Under this situation, the *shoshika* problem has been an issue in planning school recovery in the Tohoku Region and discussions on up to what scale should the schools recovered has been, in most cases, still in the air. This is rather a sensitive issue because the status of school for some communities is closely linked with survival of the community itself because local residents will require their communities to have good schools for their children. This brings these questions for school recovery: How and up to what scale should schools be rebuilt or rehabilitated? How schools should be utilized to further provide benefits for students and the whole community? Some hints to answer these questions may be found below.

Since the 1990s, around 400–500 public schools (elementary, junior high and high schools) have been closed down every year in Japan due to *shoshika* and merge of cities and towns from administrative reforms. These closure of schools does not only cause inconveniences to students and parents whose education have been temporarily disrupted and force them to travel farther from their homes to attend school in an unfamiliar environment. In the three disasters affected Tohoku Region, the total number of schools being permanently or temporary closed during 1992–2010 had already reached 574, with the majority of them being elementary schools (MEXT 2012a, b). In addition, as of 2009, some 2,254 classrooms are left unused and available for alternative use because there are not enough classes to fill the classrooms up (MEXT 2009). These numbers are expected to increase from the effects of EJET. Hints for alternative and efficient utilization of schools, in which MEXT's concept of School Centered Community Recovery and Building is based on, can been observed in MEXT Director General's notification that was released in October 1997 to encourage configuring schools with community centers/facilities, cultural facilities, sports facilities and others in the same building or same lot, so that they will have interactive functions as public facility. MEXT at this time started implementing an initiative called, "Project to Connect Abandoned School Buildings for the Future," in which it works as a matchmaker in connecting local government with interested entities such as private companies, NPOs and social welfare offices. Through this project, MEXT and the local government have been able to transform former school buildings into community centers, healthcare centers, child daycare centers, nursery homes, special needs school and office for private companies, while others have been transformed into facilities that support school activities, lifelong education programs and venue for community events. As many of the closed down schools were originally designated evacuation centers, some of these new facilities continue to function as evacuation centers or have been retrofitted to become warehouse for stockpiling and storing emergency response equipment. Some of these transformations were proven successful in benefiting the well being to community because the projects reflected the current needs of the community members. Community's familiarity of the facilities, being former schools that the community members themselves may have attended, helped them feel easy to use. Examples of successful alternation of closed schools from Iwate and Yamagata prefectures are shown in the pictures below (Fig. 5.3a, b). They both function as social welfare facility and continue to serve the school children as well as other visitors from local outside areas.

From these success stories, school recovery from EJET may be considered as an opportunity to configure schools to become multi-functional and benefit the whole community for utilization. The new school then would have a new role in facilitating interactions among community members to strengthening their ties. Planning new and creative utilization of schools by such measures should be considered and adopted before constructing new schools or permanently close down existing schools. With this said, transforming schools as multi-functional facilities may become the starting point to build a system to support school–community linkage that will allow the whole community to jointly take part in child raising and community building that are very much needed in the disaster affected areas facing the *shoshika* problem.

a

Former gymnasium is used for DRR activities and for integrated study for elementary school

Preschool – Senior Citizen's Club Joint Sports Festival

b

Former Itsukashi Elementary School Building

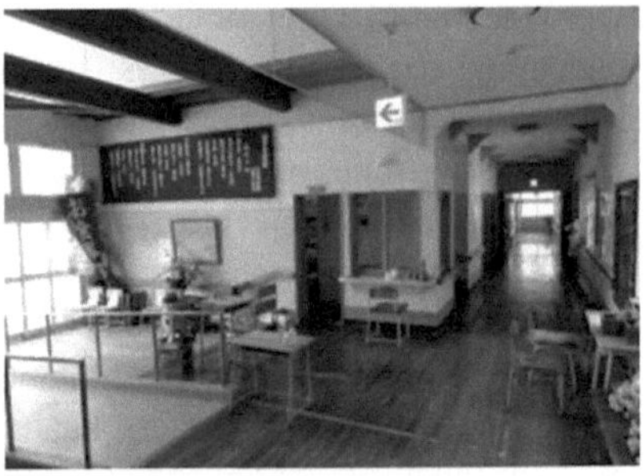

Social welfare facility opened in 2006 and operated by former School Management Committee

Fig. 5.3 (**a**) Ueda Community Disaster Management Center (former Ueda Elementary School, Sakata City, Yamagata Prefecture) Source: MEXT. (**b**) "Kibou no Oka" Social Welfare Facility (former Itsukashi Elementary School, Itsuka City, Iwate Prefecture). *Source*: Organization for urban–rural interchange revitalization

5.2.3 Shoshika *and Its Effect on Education and School Management*

The *Shoshika* problem is not merely about the structural aspect of schools, but also an issue that has been affecting school education and management and school–community linkage. The report by the Central School Council in 2000 states that communities with fewer students in schools have shown negative consequences, such as:

- Fewer opportunities for children to develop discipline in the society through challenges
- Overprotection of children by their parents
- Difficulties to pass on knowledge and experience of child raising
- Difficulties in organizing school and club activities that requires a certain number of students to conduct (such as sports festival and regional events)
- Loss of sense of competitiveness because of fewer students

Among the negative impacts above, the most crucial problem that *shoshika* has brought is the lessened opportunities for children to interact with different groups of community members. While in the past, children's social maturity was fostered through various kinds of friendship, rivalry and mutual learning through connections with different age groups in the community (Central School Council (MEXT) 2000). However, these kinds of interactions are less seen nowadays, especially for children in small schools with few classmates, lack of social adaptability, such as in ability to express oneself for large group of people, friendly competition and willingness to cooperate with others has been recognized (Hayo 2012). Overprotectiveness of the parents, who are less concerned with affairs of other children and community members, is another factor that distance their children in conversing with various people in the community (Board of Education of Ibaraki Prefecture, 1998). Another factor is the difficulty in organizing school events, such as group learning classes and sports festival, cultural festival and after school club/team activities, which has been one of the main reasons for schools to close down or be merged with other schools. These are important events in which various community members can have direct interactions with school and students (Central School Council (MEXT) 2000).

The difficulties caused by *shoshika* that affect school recovery and the longer-term community building must be met with immediate measures. A council established under the Basic Act for Measures to Cope with Society with Declining Birthrate in 2003 has been developing series of policies to suggest ways to solve the problems, but most of these are concentrated on support for parents for their child-raising. Likewise for *koreika* and population drainage in non-urban areas, the measures are focused on improving social welfare and job creation and not much has been discussed on the effects of these issues on educational contents, school management or linkage between schools and communities. Nevertheless, it is worthy to be proactive in suggesting ways to deal with this issue and possibly through educational rooted to and open to the communities. The following section will further discuss on this point by reviewing the importance of school–community linkage for school centered recovery and community building and how it can be strengthened through school education and school management.

5.3 Importance of School–Community Linkage for School Centered Recovery and Community Building

As stated previously, the effect of EJET has been greatly affecting the school–community linkage from various causes, such as closing of schools, psychological trauma of people affected by the disaster, being forced to live in unfamiliar environment and for children, studying in temporary schools away from their friends. On the other hand, the disaster has helped community members, who are experiencing the same hardships to bond, but the degree of bonding has depended on how community ties were built and existed before the disaster.

5.3.1 *Strengthening School–Community Linkages: Policies Concerning School Education and Involvement of Community*

The importance of enhancing school–community links has been marked in principle documents, including the Fundamental Law of Education (*Kyoiku Kihonho*), which was revised in 2006. Act 13 of the revised law states, "Schools, homes and community are responsible to put effort in coordinating and cooperating with each other." One principal document that supports this rational is a policy paper released in 1996 by the Central Council for Education of Ministry of Education (currently, MEXT) entitled, "Education of Japan Envisioning the 21st Century" that introduces the idea of "Open Schools" in which adults in the community take active part in educational activities to help schools, particularly in teaching the "Zest for Living" to foster the humanity aspect of children to become responsible citizens of the society (Sasaki and Suzuki 2003). This idea does not merely focus or set goals in improving educational levels, but envisions improvement of community ties through participation of the whole community in school management and education.

5.3.2 *Methods of Community Participation in School Management*

Ikeda (2001) claims that in the recent years, the school–community linkage in Japan have weakened as the following:

- Interactions between children and adults of the community have become less
- Children and adults have less opportunity to take part in community activities together
- Information and issues in the community are not shared among community members as in the past
- Due to constant transfers of teachers, continuity of educational activities is difficult

Although some of the issues above have been already raised in the negative effects of *shoshika* on education, constant transfer of schoolteachers to different schools every 3–4 years can be highlighted as a significant challenge in providing continuous and consistent education for school children. Newly appointed teachers are usually not familiar with their students and the community they work and live in, and it will take time for them to start substantial communication with the parents. Inevitably, there will be a temporary gap in the intensity of school–community relationship. Therefore, it will be practical for the community, who will most likely stay longer or permanently in the community, to fill this gap to help schools with the continuity of education by actively taking part in educational activities and school management.

In order to facilitate community integration with schools, Ikeda (2001) suggests in building an Education Community that will develop new linkages between school

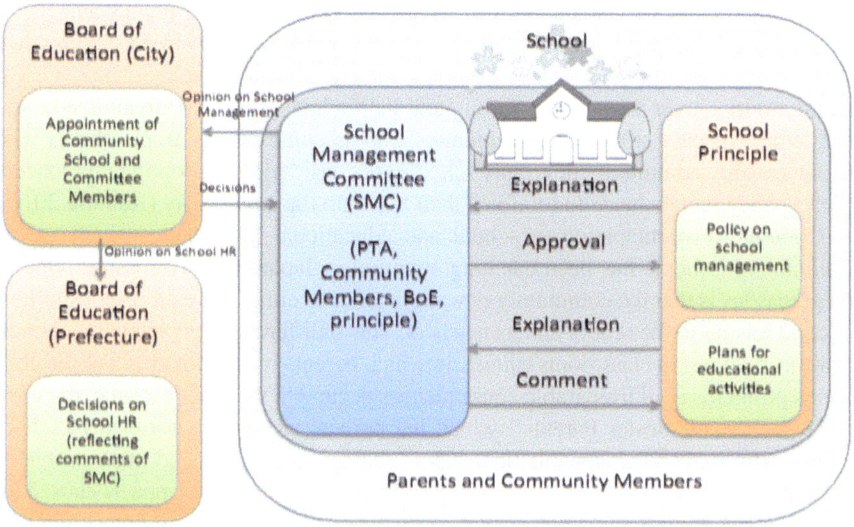

Fig. 5.4 Organizational structure of School Management Committee of Community Schools

and community by placing schools at the center of the community and allow community members windows to participate in educating their children. Education Community is a start of a shift of school from merely coordinating with its community to collaborating and integrating with the community.

From this concept arose one concrete step for the community to legally participate in school management, which is the formulation of Community Schools that was stipulated when the Law on Regional Educational Administration was revised in 2004. Community Schools are schools in which the voice of parents and community is reflected in school management through their participation in the School Management Committee. The major difference between Parents Teachers Association (PTA) other existing school committees is that while these provide comments and advice to the school principle regarding school management, the School Management Committee is authorized to have a say in the decision making of school management (Kainose 2010). If properly operated, Community Schools do not only allow community's voice to be directly reflected in school management, such as HR issues, but can also facilitate parents and other community members to better understand their roles and how they can assist the schools in their communities. Figure 5.4 shows a model organizational structure of School Management Committee that manages a Community School. A report of Central Council for Education in March 2004, expresses expectations of Community Schools taking an important role to improve communication between schools, households and community for mutual understanding and encourage communities to engage and contribute in school affairs.

An example from a school in Mitaka in Tokyo shows an additional step in the process of formulating Community Schools by combining elementary and junior high schools to form a joint school that is aimed to strengthen linkages among community, teachers and especially students of different age groups. One of the advantages of joint school is that they can maintain consistency and avoid gaps in education when students advance to higher grades. Joint school also allows the school and the community to look over the students throughout their adolescence in a streamlined manner. While success stories of Community Schools come mostly from schools with strong existing community ties, more schools are considering in adopting this concept because they feel the need to reinforce their ties with the community (Kainose 2010).

Community participation in school and educational affairs by such means as Community Schools has been catching attention in Japan, for two reasons. One of these reasons is that the community nowadays tends to either leave all responsibilities of child raising to the school and its teachers. This indifference may lead to such incidents as bullying in classrooms (*ijime*), in which parents would not aware of the children's problems until irreparable consequences occur. On the other extreme, there are the so called, "Monster Parents," which the parents are overly involved in school affairs, who make unreasonable demands to the schools. These and other problems associated with schools and students are both results of negative impacts caused by weak school–community linkage that result from insufficient communication and mutual understanding, eventually affecting community ties and trust.

As of April 2012, there are 1,183 schools designated as Community School in which the majority consists of elementary and junior high schools. In the three Tohoku Region, there are only 13 schools (Iwate 6, Miyagi 1 and Fukushima 6) that are designated as Community Schools. Although the number of Community Schools in Tohoku Region is relatively small compared to other regions in Japan, the Board of Education of Iwate Prefecture, for example, has been implementing the "Project on Promoting Community Schools in Iwate Prefecture" since 2007 and has been evaluating the progress of Community Schools in the prefecture through such reports as, Research on Coordination and Cooperation of Lifelong (Social) Education and School Education. The report suggests specific methods to link school and community through education of Community Schools (Board of Education of Iwate Prefecture 2009). MEXT recently has also proposed a new education reform policy on "Building Schools Together With Regional Community," with the purpose to strategically increase the number and regionally expand the formulation of Community Schools with the goal of 3,000 elementary and junior high schools by 2016 (Japanese Council for Community Schools 2012). As the case, although there are still few cases of Community Schools in Japan, schools and communities now better understand the benefits and opportunities that Community Schools can bring for managing schools, especially in the disaster affected Tohoku Region.

5.3.3 Exploring Different Types of Education Programs for School–Community Linkage

Community's engagement in school education is commonly to seen in the integrated study program (or *Sogo Gakushu*), which is part of the extra curricular program in Japanese school education. The contents of integrated study may include issues ranging from health, social welfare and environment/disaster management. In general, specific topics are decided by the school and its teachers. The whole aim of the Program is to let students experience and better understand the local industry, culture, history and environment and to see how adults are living and taking roles in the community. Integrated study provides school children with chances to receive "real life" education, sometimes outside of the classrooms, for the students better understand about their own community and develop a sense of responsibility as a citizen to make contribution in making their community a better place. This fosters their feeling of pride and love for their hometown, which is one of the aims of the integrated study program. Similar activities are also conducted in the regular curriculum, but would observe less involvement of the community. Informal education, such as family education or social education, is a more flexible way for community participation, but would have less involvement of schools and more likely to be conducted on an ad hoc basis. However the form, these different types of out-of-classroom education, developed after the adoption of *Yutori* education (translated as "relaxed education") in the late 1980s, provide multiple entry points for community to participate in educational activities of the school.

An example of activities conducted under the integrated study is the experience/career study program in Kamaishi City of Iwate Prefecture. Because the major local industry in Kamaishi is fishery, local schools including Toni Elementary and Junior High Schools have been cooperating with the Fishermen's Cooperative Association to carry out salmon hatchery, release and food processing experience programs (Suda 2012). Since students usually do not have much connection with the fishermen or have actually seen how salmon is processed, this becomes a precious experience for school children to better understand how their local economy has been supported. Some may even consider taking such profession in the future. Although the classes were temporarily disrupted by EJET, the schools in Toni District have resumed this program since late 2011. In the next academic school year, Toni Elementary School mentioned above, also plans to incorporate Recovery Education in their integrated study in which school children will take part in volunteering activities and interviewing construction workers doing recovery work (Document).

Although not solely under the jurisdiction of MEXT, a useful reference that shows community participation in education is the program under the Children After School Plan, in which MEXT and Ministry of Health, Labour and Welfare (MHLW) jointly administer. The content of the program varies from school to school, but commonly, students take part in history and culture studies, games, art, sports and disaster/safety management activities. Community involvement and frequency to the programs also differs, but implementation in most cases requires assistance from community volunteers who own experience and knowledge that schoolteachers do not have. The Children After School Plan provides a win–win situation in which children are able to interact with people of different age groups to learn about life. The adults are also given vitality and a sense of purpose and a role in making contributions to the community. It is important to note that these education programs also involve community members who do not have children in schools and do not have direct linkages with the schools.

5.3.4 ESD: An Approach for Community Linkage in Education

One innovative approach in conducting integrated study for providing schools and communities with sustainable education is the concept, Education for Sustainable Development (ESD) that the Japanese Government had proposed in the Decade of Education for Sustainable Development (2005–2015) at the World Summit on Sustainable Development in Johannesburg in 2002. The significance of ESD is that it encourages teachers to "go beyond the school walls, establish links with community and professional organizations and institutions, and promote educational activities with the support of a broader partnership framework" (Oikawa 2012). Tanaka (2003) raises some of the featured themes in ESD, which are:

- Environmental education on the ecosystem and environmental conservation,
- Education on development issues such as overpopulation, poverty and health,
- Education based on peace, human rights, democracy and coexistence.

ESD focuses not merely on teaching the technical features of these topics, but rather, on promoting these common social issues through awareness raising and assist the community to work together in partnerships to solve these problems (Nitta 2002). In addition, ESD places importance on the participatory approach in creating a sustainable society, which children can explore and find "links between themselves and adults, their local regions and the world" (Abe 2002). As issues like environment management and DRR are topics that are not confined in schools, but also at homes and in the surrounding environment, the concept of ESD helps schools reach out and make linkages with the community and the environment around them. As ESD nourishes children's critical thinking, systematic thinking, information collection and analysis, decision making and especially the ability to communicate, ESD can be an effective tool to strengthen the linkage or partnership with the community (Oikawa 2012). This will be a major force that will develop mutual support to develop capacity in DRR as well as propelling the recovery process from disaster situation.

5.3.5 Function of Japanese School as Community DRR Hub

In Japan, school and community usually have existing linkage because in many cases, schools are designated by the city government as evacuation centers during emergencies and as hubs for providing DRR education. Two important laws that concern adopting schools as evacuation centers are the Disaster Countermeasures Basic Act and the Disaster Relief Act. Both laws encourage public facilities such as schools and community centers to be designated as evacuation centers, provided that they are earthquake proofed and owning proper facility, such as toilets and water supply facility and equipment including communication radios and TVs. According to the National Institute for Educational Research (NIER), around 89.3 % or 30,513 schools in Japan are designated as evacuation centers (as of May

2011). During EJET, 622 schools, including non-designated schools were used as evacuation centers. Out of these, many of the closed down schools also functioned as evacuation centers. A survey conducted by MEXT shows that out of 2,443 closed schools, 1,262 or 51.7 % carries on their important function as temporary evacuation centers (MEXT 2011a, b, c, d).

As for school's function as an effective hub for DRR, a report by the Central Council for Education in March 2012 provides an example from the Great Hanshin Awaji Earthquake (GHAE) that shows how schools, particularly those that already had close linkages with surrounding communities before the disaster, operated smoother as evacuation center. The report continues to claim that disaster preparedness gained through DRR education that promotes community ties, will empower communities with not only knowledge, but also close bonds that could stand up to hardship during disasters and take initiatives in the rehabilitation process.

In practice, this can be observed in the voluntary/community based Disaster Welfare Community or BOKOMI that was established and mainstreamed in the city of Kobe in 1997 in which school–community linkages that existed before the earthquake disaster was utilized (Matsuoka 2011). BOKOMIs were formulated in every elementary school districts because welfare community groups were already established in each elementary school, making integration of DRR factors into these existing groups easier. BOKOMI holds an ideal combination of functions to support the communities during emergency and DRR, as well as supporting welfare activities that are needed by the aging communities during non-emergency times (Kaneyoshi 2004).

An example from EJET proved likewise, in that school, which had the Regional School Support Center established before the disaster also had less issues in evacuation operation (Central School Council (MEXT) 2012). The Regional School Support Center is a system established in 2008, mostly for junior high schools, in which community volunteers support schools with such activities as helping classroom and afterschool activities, operation of school library, maintenance of school environment and assisting DRR activities, such as evacuation drills.

5.3.6 School–Community Linkages Through DRR Education

In 2006–2007, the United Nations International Strategy for Disaster Reduction (UNISDR) promoted with the campaign "Disaster Risk Reduction Begins at School." Although the campaign underlined the importance of conducting DRR education in schools, it also emphasized the significance of community participation in DRR education for the sustainable development of the whole society. As the case, DRR education may begin at schools, but not confined in schools. Students must learn to protect themselves not only in schools, but also at their homes and also on their way to schools.

Referred by UNISDR as model city for DRR education with effective school–community coordination, Saijo City in Ehime Prefecture has been conducting the "12-year old Education Project" since 2006, in which not only the students, but parents, community association members and local government officials jointly take part in DRR education activities, such as town watching and mountain watching to raise community awareness and to understand the roles of each stakeholders (Shiwaku and Fernandez 2011). The students who participate in this program are able to convey what they have learned about DRR at school and through DRR activities to their family members in the Family DRR Meeting conducted at their homes. As most parents would not ignore what their children have to say, the adults are encouraged to listen and participate in DRR drills, prepare emergency backpacks and become aware of the possible dangers around their homes. The Family DRR Meeting also allows children to ask their parents and grandparents about past disasters and ideas on taking necessary measures. This kind of co-learning within homes and communities provides opportunities and encouragement for the whole community to get involved to work on a common issue that enriches community ties. The results of these activities are reported at the annual Children's DRR Summit with participation of parents and other community members (City of Saijo 2012).

From the Saijo model, building the culture of DRR indeed seems more effective when the advocacy is targeted toward school and student rather than their parents as adults are usually busy with work and they are selective in their interests to seriously consider behavior modifications. Katada (2012) shares his experience in promoting DRR education, shifting the original target audience who were adults to students to become advocators of DRR. The idea takes advantage of parents' nature of them listening more to their child instead of others outside of the family. Adults in the community usually are concerned of what their children are being taught in schools, students can become promoters of DRR through interactions with their family and community (Katada 2012). Traditionally, DRR knowledge was passed on to other family members through these kinds of daily activity at homes (Takeuchi et al. 2011), so the approach is somewhat of a revisit to how DRR education was conducted in the past. The approach in targeting DRR education for students for dissemination proved to be also effective in Kamaishi City during EJET. The school children were not only able to initiate their own evacuation, but also were able to encourage or help other community members to escape with them, producing a sense of solidarity in the whole community to better cope with the disaster. Table 5.1 shows different components of school community linkages.

5.4 Building School Centered Disaster Resilient Communities

With the issues on low birth aging society in Japan, community participation in school management and education to strengthen school–community linkage and trust, MEXT has come up with the concept of School Centered Community Building to aim for early recovery of the affected areas of Tohoku Region through the recovery of schools.

Table 5.1 Components for enhancing School–Community Linkage for School Centered Recovery and Community building

Component	Contents and expected effects
School management	Consideration for building Community Schools in which decisions by community members take part in school management, including decisions on human resources, school activities and policies
	While schoolteachers are transferred to other schools in 3–4 years, community can continuously be involved in their community's school management. It also raises common interest of the community to educate and raise children
Education	Strengthen community participation and utilize community members in integrated study programs and after school programs by incorporating concepts such as ESD
	School children will be able to receive "real life" education in the context of their community and will have better understanding of their community. Community members will be able to build a sense of having roles in contributing to their community
DRR activities	Encourage school–community cooperation in DRR activities/education and build common understanding that school is a safe DRR hub
	Schools that are designated as evacuation place will be strengthened as DRR hub
	Schoolchildren will be able to build DRR capacity not only in schools, but also outside of schools and become advocator for DRR

5.4.1 *MEXT Concept for School Centered Community Building*

On October 11, 2011, then MEXT Minister, Masaharu Nakagawa, announced the concept for School Centered Community Building addressing the affected people of the East Japan Earthquake and Tsunami. On the same day, Vice Minister Takashi Kii followed by notifying 1 municipal and 15 prefectural Board of Education regarding rehabilitation and recovery of public school facilities in the affected Tohoku Region. Both announcements referred to school as significant symbol and central hub of the regional community and that it is quick recovery of schools is essential in providing conditions for displaced people to return to their communities and strengthen community ties that have been weakened from EJET. Through school recovery, MEXT is hopeful that it will become one of the key elements that will contribute to the whole recovery process of the affected areas. The concept originates from the many discussions among such groups as, Commission on Improving School Facilities Based on Damages Caused by EJET, organized by MEXT and Working Group on Concept Building for Recovery from EJET, organized by Cabinet Secretariat among others.

Special provisions are provided to EJET affected schools to facilitate quick recovery and efficient use of recovery budget by simplifying administrative procedures, such as adopting a system that allows local governments to start with the project before submission of the application or receiving official field survey from MEXT. In addition, although in principle, the budget should be used to rehabilitate the schools back to their original state, in case it is not possible for schools to be reconstructed in their original place, due to such reason as land subsidence caused by earthquake, local BoE is able to use the recovery budget to relocate the schools to safer lands. Additional budget may be available to strengthen DRR function of the schools (MEXT 2011a, b, c, d).

The third supplementary budget, inclusive of recovery aid grant, allocated to newly construct, rehabilitate or relocate public schools totaled around 47.6 billion JPY (about 476 million USD) in the 2011 Japanese Fiscal Year (JFY), in 2012 JFY the budget allocated mostly to conduct necessary relocation schools amounted up to 14.7 billion JPY (about 147 million USD) and for 2013 JFY, the budget plan totaled to 21.2 billion JPY (about 212 million USD). The budget utilizes both general account budget and special account budget for recovery (MEXT 2011a, b, c, d, 2012a, b, 2013). As stipulated in the Act 3 of Law Concerning National Liability for Expenditure on Rehabilitation of Public School Facilities, the national government will bare 2/3 of the total construction cost to rehabilitate affected schools back to their original state and also for constructing temporary school buildings. In addition, there is recovery aid grant that is administered by prefectural government and proposed by city governments that can be applied for to initiate school recovery projects. For example, Kamaishi City had applied for 3.7 million JPY from this grant to survey and design the new Toni ES/JHS and also to rehabilitate the Toni CC that is planned to be built together with the schools.

Table 5.2 is a list of memorandums and notices that MEXT has released in 2011 regarding school rehabilitation and recovery. As seen from the list, measures currently concentrate on recovering the physical infrastructure, but already some issues are arising. For example, the expenditure rate in JFY 2011 only reached up to 27.8 % due to such issues as delays in the overall recovery plan of local governments, difficulties in selecting new and safe plot for schools and from differences with original assessment of damages to schools. Coordination among relevant departments is also required, because community building inevitably goes beyond jurisdiction and responsibilities of the Board of Education (BoE). There are also gaps in expectations of BoE, schools and the communities because there has not been a systematic way to facilitate stakeholder dialogues to come up with a consensus. These are some of the issues that must be solved in the school recovery process.

5.4.2 *Components of School Centered Recovery and Community Building*

MEXT's concept of School Centered Community Building is based on four components under two main objectives in rehabilitating and recovering the affected school facilities. One of the objectives is to rehabilitate or reconstruct the school building affected by the EJET to resume normal educational activities. The other objective is to recover schools for the purpose to realize a longer plan to make schools hubs or centers of the regional community. In order to accomplish both objectives, strengthening the structural features of the building from future disasters, especially earthquakes, is of all things required with urgency. In addition, schools need to be

Table 5.2 List of notification regarding rehabilitation of public school facilities that was affected by the East Japan Earthquake and Tsunami

(Rehabilitation process)
1. "Regarding early implementation of rehabilitation of public school facilities" Memorandum on August 10, 2011

(Advance start of construction)
2. "Implementation of rehabilitation work for early resumption of school education" Notice 23-2 on April 4, 2011
3. "Regarding early rehabilitation of school facilities damaged by East Japan Earthquake and Tsunami in 2011, Memorandum on 15 and 17 March, 2011.

(Simplification of application procedure)
4. "Notice regarding implementation design of building rehabilitation/repair, construction management of new construction – repair – rehabilitation and rehabilitation of land of East Japan Earthquake and Tsunami", Memorandum on August 1, 2011
5. "Regarding processing of national budget support to rehabilitate school facilities from disasters of East Japan Earthquake and Tsunami", Notice 23-15 Department of Facilities Planning and Administration on June 7, 2011.
6. and 7. "Regarding procedures to survey budget to rehabilitate public school facilities under jurisdiction of MEXT from the East Japan Earthquake and Tsunami" Notice 23-13 Department of Facilities Planning and administration and 15 on June 7, 2011
8. "Regarding damage situation report, plan of national budget support operations and handling of photographs of disaster situation from the field" Notice 23-8 of Department of Facilities Planning and Administration on May 20, 2011.

(Tsunami Disaster)
9. "Handing of rehabilitation of public school facilities affected by tsunami" Memorandum on September 2, 2011.
10. "Partial revision to guideline in disposing hazardous waste utilizing national budget support" Memorandum on August 19, 2011.

(Temporary school buildings)
11. "Procedure to survey budget to rehabilitate public school facilities under jurisdiction of MEXT from East Japan Earthquake and Tsunami" Notice 23-30 of Department of Facilities Planning and Administration on September 2, 2011.

(Management of soil)
12. "Addendum to the guideline for budget support to rehabilitate public school buildings and handling of guideline to survey budget to rehabilitate public school facilities under jurisdiction of MEXT" Noteice 23-21 and 23 of Department of Facilities Planning and Administration on June 20, 2011.
13. "Strategy to reduce air dose in school grounds of schools based on field survey", Memorandum on May 11, 2011.

retrofitted to function better as temporary evacuation centers with basic infrastructures such as storage for emergency provisions, adequate number of toilets, communication system and equipment to provide electricity and water during blackouts and suspension of water supply. For the sustainable use of the facility and for environmental education, ecofriendly features should also be improved, such as by installing solar panels. Lastly, rebuilding school to become a multi-functional public facility is encouraged so that the functionality for public services, including DRR functions can be strengthened for schools to become central hubs for community recovery and building.

Among four components, rebuilding or retrofitting school to become a multi-functional public facility is the key in materializing the concept of School Centered

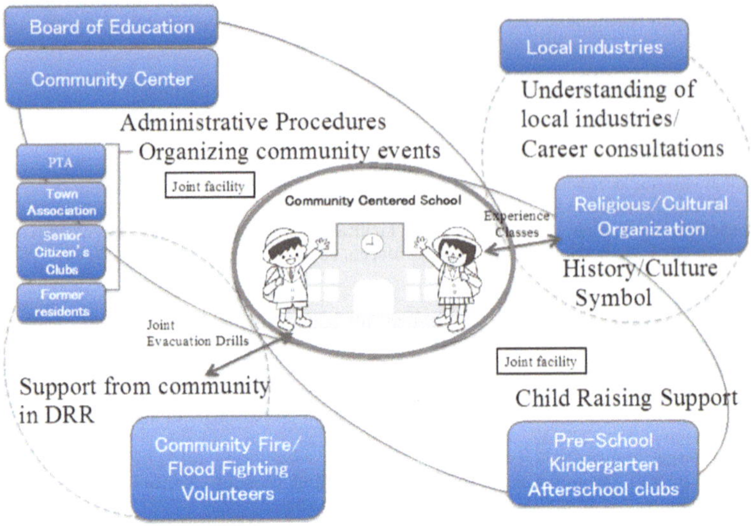

Fig. 5.5 Image of multi-functional school facility

Recovery and Community Building. The idea of combining or building multiple public facilities adjacent with each other is the idea promoted by MEXT for schools affected by the EJET. MEXT rationalize that the concept will, (1) Provide efficient use of land, especially in regions with limited plot land, (2) Provide an educational environment in which the whole community can protect and support schools and children and (3) Strengthen community ties from interactions among different generation of people. An idea of a school rebuilt or retrofitted as multi-functional public facility is shown in Fig. 5.5.

The figure shows several of the ordinary public facilities and stakeholders seen in rural townships in Japan. These facilities or its functions can be either considered to be incorporated into the school or arranged to have close collaboration to assist the school management and educational activities.

- Community Center is a public facility that provides social education for the community. Stipulated under the Social Education Law Act 20, it promotes lifelong education through educational, cultural and other recreational activities reflecting the local context and needs. It is also a place where town associations and local resident's clubs (such as Senior Citizen's Club and Women's Union) organize regular meetings. The center may also organize community events, give health consultations and disseminate information by such means as community newspaper. If the community center is incorporated with school, both can mutually support and benefit in providing various kinds of education for the whole community. Community Center is under the jurisdiction of the Board of Education.
- Kindergarten, Preschools and After School Clubs play an important part in support parents in their child raising. This is because it is common to see dual income families not being able to take care of the children after school because both

parents will still be at work. As noted in the previous section, the Children after School Plan implemented by the two ministries, MEXT and MHLW, promotes strengthening of these facilities as a measure to cope with low birth society. Kindergarten, preschools and After School Clubs are normally well coordinated with elementary schools, so integration of these facilities with school would enhance the convenience of the community. Preschools and After School Clubs are usually under the jurisdiction of Child's Division of the Health and Social Welfare Department and kindergarten is under the Board of Education.

- Local industries and religious/cultural organizations are also essential in providing school children with understanding and pride for the communities they live in. Through such channels as integrated study programs, students may receive valuable "real life" experience and career education in which schoolteachers, in most cases, are not able to provide. It is also important to note that these industries and organizations also receive benefits from their involvement because they will be able to gain a sense of making a social contribution and support to the children who are the future of their respective communities.
- Community Fire/Flood Fighting Volunteers (Syobo-dan and Suibo-dan) plays a vital part for maintaining safety of the community and disaster management. They can also contribute to DRR education and activities conducted in schools through DRR demonstrations or lectures. Some of these volunteers may be family members of students, which will make them proud of their family. On the other hand, these volunteers will also benefit as they will be able to better grasp the school and its children's profile and needs that will be helpful when assisting them during emergencies. Both fire/flood fighting volunteers are commonly under the jurisdiction of the local fire department.

As seen from the descriptions above, the key for successful implementation of this concept is the mutual benefit of both school and community. Another element is the coordination mechanism of related local government office departments responsible for these organizations and stakeholders. For this, the functional roles of these stakeholders may be clarified in the whole concept. For example, Community Centers with their affiliation with the local government can take the institutional and budgetary related functions. The local industries and cultural organizations can take the economical and cultural role that will provide regional uniqueness to the school. Kindergarten and Preschools can take the role in social support by helping parents raise their children. The Community Fire/Flood Fighting Volunteers can take the function as DRR advocates in ensuring safety of the school during and out of emergency times.

5.4.3 *Analysis of Benefits and Challenges of School Centered Recovery and Community Building*

With this image of making school a multi-functional facility as pictured in Fig. 5.6, a SWOT (**S**trengths, **W**eaknesses, **O**pportunities, **T**hreats) analysis is conducted to root out the benefits and challenges in realizing the concept of Building School Centered Disaster Resilient Community (Fig. 5.6).

	Internal Factors	External Factors
Positive Factors	**Strength (S)**	**Opportunity (O)**
	• Strong link of community achieved strengthened school – community linkage • Students and community have pride in their community's history, culture and education • Community takes part in school management and educational activities • DRR capacity is strengthened through joint school - community DRR activities • New school contributes to interaction among community members	• Coordinated recovery improves infrastructures and environment around the new school • Opportunities to receive funds and assistance for school recovery that benefit the whole community • Number of students increase as a result of building attractive school/multi-functional facility
Negative Factors	**Weakness (W)**	**Threats (T)**
	• Programs to encourage school - community linkage that benefits everyone need to be continuously planned and implemented • Further strengthening of DRR plans and capacity needed for through school – community joint activities • Some students still need psychiatric care • Need to ensure making multi-functional facility does not burden the school management	• Decrease of population due to low birth rate and aging may further downscale the school • Difficult to predict the recovery progress of the overall city recovery to realize school centred community building

Fig. 5.6 SWOT analysis for School Centered Community building

Through the SWOT analysis above, the following can be said about the benefits and the challenges that must be either strengthened or taken measures to. The internal factors are features that exist within the subject community while external factors are elements that are outside of the community that affect the community.

5.4.3.1 Internal Factors

- Strength: School as a multi-functional facility will become a hub for school–community interaction, specifically by community participation in school management and education programs to rebuild and strengthen school–community ties. Education that is deeply rooted with the region will nourish a sense of pride for the whole community about their hometown.
- Weakness: A system to sustainably implement activities for school–community interactions and DRR activities must be established in order to maintain the benefits. Retaining continuity for these activities may not easy, especially considering the psychiatric conditions of school children as well as adults who were affected by EJET. In addition, multi-functional facility may complicate management/security issues, therefore, roles and responsibilities must be clarified at early stage of implementation.

5.4.3.2 External Factors

- Opportunities: If the overall recovery is well coordinated, infrastructure and the environment surrounding the school will be improved, benefitting both school and the community. When the benefits become clear, displaced people will start returning to their communities and even people outside of the community might become eager to move in. This will become a positive factor that will increase the population of the community, that will revitalize the school and community.
- Threats: As mentioned repeatedly in this paper, the population decline for rural communities and disaster affected communities is a problem that individual community by themselves cannot solve. For this problem, outside help, such as from NPOs, industries and experts, might be required to take effective measures. Additionally, as it is difficult to predict the recovery process, especially after mega disasters like EJET, building a system to monitor and evaluate each step of the recovery process in order to plan the next steps is required.

As construction of the school facility by itself will not sufficient to sustainably and efficiently usage, actions must be taken according to analysis that looks into the positive and negative factors for school recovery to respond to the possible bottlenecks and to ensure long benefits for the community.

5.5 Conclusion

This chapter has highlighted the importance of community ties and trust in disaster response and recovery as well as in the longer term community building. It further argued that in order to restore the weakened or lost community ties due to EJET, school recovery may be one of the most effective ways to encourage and facilitate the community recovery process. An example of post-disaster recovery of a school district in Miami-Dade County in South Florida that was devastated by the Category 5 Hurricane Andrew in 1992, demonstrates how recovery of school (a) was important in reestablishing a sense of community, (b) gave schools role as focal points in which community could unite, (c) created sense of identity for students and faculty and (d) gave schools role as social and medical support centers (Provenzo and Fradd 1995). Particularly for Japanese school districts, which are usually confined in a relatively small area where students can walk to their schools, have also taken important roles in building unity and linkage within communities as different generations have all received education there (Nishida referred in Sakagawa 2004). As the case, despite the fact that many affected schools are now being rehabilitated or newly built, and the students are still displaced in different schools, the sense of school community, extending beyond students and teachers to the whole community, persisting in the affected communities. This implies that prompt school recovery can become a key factor in rebuilding and strengthening community ties that can facilitate the overall community recovery.

On the other hand, the challenges for school recovery were also described in this chapter. Population drainage caused by the immediate effect of EJET and from the chronic problem that has existed before EJET, poses a major challenge in school recovery and for the continuity of school operation. In the prolonged recovery from EJET, the phenomena of low birth and aging population might be accelerated with the working age group leaving their hometowns to look for better opportunities, leaving the elderly people behind. With the dwindling student number, some even argue whether schools should be rebuilt as it was before EJET, in questioning the efficiency of investing in new school buildings. Conversely, in a group discussion held in the "Workshop on Enhancing Disaster Resilience of Education Sector and Communities" organized by Kyoto University on August 2012, closing or scaling down schools may become a community survival issue, because the decreasing number of children will most likely to continue, further aging the communities (Shaw 2012). As the recovery and community building process in Tohoku is still expected to become a long process, it is essential that all age groups, especially the younger generation, to stay or return to the communities to work together in realizing the future visions for recovery.

In overcoming these bottlenecks, MEXT's concept of "School Centered Community Building" was introduced as an applicable option. The concept proposes in recovering affected schools by physically incorporating multiple public facilities to the school building that will make the school to become a platform to facilitate interaction among community members. It is also aimed to ensure the efficiency of recovery investment, by encompassing the possibility of decreasing student number in the future. Another important factor is reinforcing its function as DRR/evacuation center and to raise community awareness to seek safety in designated schools. In this way, the new school can meet the need to be a catalyst in strengthening community ties, provide safe environment for the community and at the same time, be able to adapt to future demands required as a core public facility.

Aside from the structural aspect perhaps the more important part of school recovery is developing a mechanism in which communities can continuously be linked with their schools. Past and ongoing efforts in linking schools with their communities were introduced with such concepts as "Community School" that provides opportunities for communities to be involved in school management. Also through educational activities, communities have various windows for participating in school operation, including integrated study in which the whole community can contribute with their experience in providing children with "real life" education. In this context, the concept of Education for Sustainable Development (ESD), mentioned in this chapter, can be a tangible option in utilizing school education as a tool to link the school with its community and incorporate regional education, such as in learning about local heritage and culture, to develop pride and love for hometown of students and their community (Oikawa 2012). In this way, education can become an effectual tool to raise consciousness of all community members, including the students, to think about recovery, community building and prepare for future disasters and be responsible in realizing these goals of the community.

Some of the success stories in the Tohoku Region have proven that the strong ties and trust within families, friends and communities that existed before the EJET have helped people to make the right response decisions by putting the teaching of *Tsunami Tendenko*, a traditional self-help teaching that holds individuals responsible in protecting one's own life, into actual practice. (Yamashita 2011, Katada 2012). Likewise, for post-disaster recovery, communities that have strong social networks will experience faster recovery, as it is likely that they will have access to more information, tools and assistance for recovery (Aldrich 2012). While many of these ideas raised in this chapter existed before EJET and not entirely new, school centered recovery can become a powerful option to renew the entire school system, while providing an opportunity and a reason for dispersed people to return to their communities and contribute to the recovery and community building process.

Acknowledgements The authors acknowledge the financial support of CWS and MERCY Malaysia to conduct fieldwork in Tohoku disaster affected areas.

References

Abe O (2002) Establishing a sustainable future. Kikan Erukore-da, #12, pp 1–4

Aldrich D (2012) Building Resilience: Social Capital in Post-Disaster Recovery. The University of Chicago Press

Board of Education of Iwate Prefecture (2009) Research on coordination and cooperation of lifelong (social) education and school education – research report lifelong education in Iwate

Cabinet Office, Government of Japan (2011) Disaster Countermeasures Basic Act and the Disaster Relief Act 2011

Central School Council (MEXT) (2000) Low birth society and education. Council Report on 1 April 2000

Central School Council (MEXT) (2012) Report on decision on plan to facilitate school safety

City of Saijo (2012) Disaster management plan with participation of all community members: for you after 30 years. Crisis Management Division, Administrative Department, Saijo City, p 193

Guha-Sapir D, Vos F, Below R, with Ponserre S. Annual Disaster Statistical Review 2011: The Numbers and Trends. Brussels: CRED; 2012

Haghebaert B (2007) Working with vulnerable communities to assess and reduce disaster risk. In: Humanitarian practice network at the Overseas Development Institute No. 38, pp 15–16

Hayo M (2012) Sustaining proper school size by way of school merger face with long-rage population decline and restructuring of local school system. In: National Institute for Educational Policy Research Bulleting No. 141, March 2012

Ibaraki Prefecture Board of Education (1998) Special article on education in low birth – aging society. http://www.pref.ibaraki.jp/forum/forum03/f3-4.pdf. Accessed 26 Nov 2013

Ikeda H (2001) Education community handbook: seeking for linkages and collaboration of community and schools. Kaiho

Ishiwatari M (2012) Government Roles in Community-Based Disaster Risk Management. In Shaw, R. (Eds) Community-Based Disaster Risk Management. Emerald Group Publishing Ltd. pp 21–27

Japanese Council for Community Schools (2012) Council Bulletin (10 July 2012). http://www1.ocn.ne.jp/~jccs2011/newsletter.html

Kainose S (2010) Establishing Joint Elementary and Junior High Community School: challenges of Superintendent of Mitaka Board of Education. Poplar Publishing

Kaneyoshi T (2004) 10 years of recovery – the fight of Kobe: emergency measures of local governments and schools. pp 156–159. Nihon Keizai Shinbun Publishing

Katada T (2012) Education that saves lives: lessons learned from 3.11 in Kamaishi. pp 38–42 and 98–104. PHP Inc.

Kyodo News Agency (2012) Article on school recovery in Tohoku region on June 22, 2012

Matsuoka Y (2011) Partnership Between City Government and Community-Based Disaster Prevention Organizations in Kobe, Japan. *In* Shaw R. (Eds) Community Based Disaster Risk Reduction. Emerald Group Publishing Ltd. p.151–184

Ministry of Education, Culture, Sports, Science and Technology (MEXT), Japan (2013) Community school that connects school – family – community: Division of Afterschool Plan Development/ Coordination MEXT/MHLW. http://manabi-mirai.mext.go.jp/. Accessed 26 July 2013

Ministry of Education, Culture, Sports, Science and Technology (MEXT), Japan (2009) Document on unused classrooms (dated 1 May 2009). http://www.mext.go.jp/component/a_menu/education/detail/__icsFiles/afieldfile/2009/12/04/1286096_2.pdf. Accessed 26 July 2013

MEXT (2011a) Situation on damages to school facilites caused by East Japan Earthquake and Tsunami. http://www.mext.go.jp/b_menu/shingi/chousa/shisetu/017/shiryo/__icsFiles/afieldfile/2011/06/28/1307121_1.pdf. Accessed 26 Nov 2013

MEXT (2011b) Reconstruction and recovery of public school facilities affected by the East Japan Earthquake and Tsunami. http://www.mext.go.jp/b_menu/houdou/23/10/attach/1311986.htm. Accessed 26 Nov 2013

Ministry of Education, Culture, Sports, Science and Technology (MEXT), Japan (2011c) Press conference on 3 and 16 September 2011. http://www.mext.go.jp/b_menu/houdou/22/09/1297186.htm and http://www.mext.go.jp/b_menu/houdou/23/09/1311255.htm. Accessed 12 July 2012

MEXT (2011d) Notification regarding administrative procedures for rehabilitation of public school facilities affected by disaster on March 13 and 17, 2011. http://www.mext.go.jp/a_menu/saigaijohou/syousai/1. Accessed on 26 Nov 2013

Ministry of Education, Culture, Sports, Science and Technology (MEXT), Japan (2012a) Information on disaster situation of the East Japan Earthquake and Tsunami, 202th report

Ministry of Education, Culture, Sports, Science and Technology (MEXT), Japan (2012b) Press Release on number of closed down educational facility and their usage (on 14 September 2012). http://www.mext.go.jp/b_menu/houdou/24/09/__icsFiles/afieldfile/2012/09/14/1325788_1.pdf. Accessed 26 July 2013

Ministry of Land, Infrastructure, Transport and Tourism (2011) MLIT white paper 2011, Chapter 1. Available at http://www.mlit.go.jp/hakusyo/mlit/h23/hakusho/h24/html/n1111000.html. Accessed 26 July 2013

National Institute for Educational Research (2011) Result of factual survey on DRR functions of school facilities. Available at http://www.nier.go.jp/shisetsu/pdf/bousaikinou2011.pdf. Accessed 26 July 2013

National Institute of Population and Social Security Research (2011) Prediction of population by prefecture in Japan (as of 2006). Accessed on November 26, 2013. http://www.ipss.go.jp/ppfuken/j/fuken2007/t-page.asp

Nitta K (2002) Using environmental education to create a sustainable society: general education for a sustainable society, Centre for Educating Global Citizens and the greatest problems facing environmental education: globalization and unsustainable societies. The Japanese Society of Environmental Education, Environmental Education, #22, pp 15–25

Oikawa Y (2012) Education for Sustainable Development and Its Implications to Recovery Process in Kesennuma. In East Japan Tsunami Earthquake, Shaw R. and Takeuchi Y. (ed). Research Publishing. pp 165–186

Provenzo E, Fradd S (1995) Hurricane Andrew, the Public Schools and the Rebuilding of Community. State University of New York Press

Reconstruction Agency Japan (2012) Announcement on July 11 2012. http://www.reconstruction.go.jp/topics/120711hinansya.pdf. Accessed 7 July 2012

Sakagawa S (2004) School as center of the local community. Kokon Shion Publishers, pp 43–44. Tokyo, Japan

Sasaki H, Suzuki M (2003) Social Education and School. Gakubunsha Collaboration among schools, families and communities, and social education (55th series on social education in Japan). Japan Society of Social Education

Shaw R, Shiwaku K, Takeuchi Y (2011) Disaster Education, Emerald Publisher, UK. pp 162

Shaw R, Takeuchi Y (2012) East Japan Earthquake and Tsunami: Evacuation, Communication, Education and Voluntarism. Research Publisher, Singapore. pp 280

Shaw R (2012) Workshop on enhancing disaster resilience of education sector and communities. Kyoto University's Graduate School of Global Environmental Studies

Shiwaku K, Fernandez G (2011) Disaster education. In: Shaw R, Shiwaku K, Takeuchi Y (eds) Innovative approaches in disaster education. pp 120–122

Statistics Bureau of Japan (2101) Population and Households Of Japan 2010

Suda Y (2012) Linking human behavior, participatory mapping community infrastructure for disaster risk reduction: approaches in Shisho and Kamaishi. Master's Thesis for Kyoto University's Graduate School of Global Environmental Studies

Takeuchi Y, Mulyasari F and Shaw R (2011) Roles of Family and Community in Disaster Education. *In* Shaw R., Shiwaku K. and Takeuchi Y. (Eds) Disaster Education. Emerald Group Publishing Ltd. pp 77–94

Tanaka H (2003) What is education for sustainable development? – Ideas for preparation (25 April 2003). Available at http://www.rikkyo.ne.jp/~htanaka/03/ESD01.html. Accessed 26 July 2013

Tohoku Bureau of Economy, Trade and Industry (2012) Situation of rehabilitation/recovery of Tohoku region and initiatives of the Tohoku Bureau of economy, trade and industry. Available at http://www.meti.go.jp/press/2012/04/20120413007/20120413007-4.pdf. Accessed 26 July 2013

UNCRD Disaster Management Planning Hyogo Office (2003) From Disaster to Community Development: The Kobe Expereince, Tsunehiro Y., Goda K., Shaw R. (eds). pp 18–19 and pp 21–25

Yamashita F (2011) Tsunami Tendenko: Tsunami history of modern Japan. pp 231–235

Chapter 6
Building Disaster Resilient Community Through Healthcare Networking

Kenji Isayama and Rajib Shaw

Abstract Experiences a disaster can be traumatic, and it can affect people not only physically but also psychologically, socially, and economically. Damage to the victims of a disaster is not only physical but can also have a significant psychological component. Recovery is not just about the rebuilding the infrastructure in an affected area but also concerns rehabilitation of individuals and the process of rebuilding their lives. Physical and mental healthcare are needed during the reconstruction/rehabilitation phase after a disaster as well as during daily life. However, "earthquake-related deaths" were reported due to overwork and deterioration from chronic diseases after a disaster. There is room for improvement in healthcare-related networking. Cuba's medical system is recognized as a good example of a network of daily medical care and disaster medicine. This system also helps community-coping capabilities when a disaster occurs. In contract, as healthcare-related approach, the World Health Organization (WHO) has proposed "Health Promotion". "Health Japan 21" has also proposed in Japan. These proposed goals are expected to be "Promotion of health in communities". These promotion activities highlight cooperation between individuals, families, communities, and healthcare-related organizations. It is necessary to maintain stronger bonds among communities. These bonds need to be utilized for the development of the city and for reconstruction after a disaster. Previous studies suggest that primary prevention is the most effective means of reducing earthquake casualties. However, it is difficult to maintain awareness of disaster prevention because disaster memories of victims are only forgotten gradually. Japan is expected to have significant earthquakes in the near

K. Isayama
Department of Medical Science and Technology, Hiroshima International University, Hiroshima, Japan

R. Shaw (✉)
Graduate School of Global Environmental Studies, Kyoto University, Sakyo-ku, Yoshida-Honmachi, 606-8501 Kyoto, Japan
e-mail: shaw.rajib.5u@kyoto-u.ac.jp

R. Shaw (ed.), *Community Practices for Disaster Risk Reduction in Japan*, Disaster Risk Reduction, DOI 10.1007/978-4-431-54246-9_6, © Springer Japan 2014

future, and a "super-aging" society essentially makes the society more vulnerable. Therefore, preparedness for future mega disasters is essential. Disaster prevention measures will maintain awareness for disaster preparedness through activities such as festivals and healthcare, and these are most important for activating areas of organizational activity. Strengthening the healthcare network as an indirect disaster preventive measure will play an important role when a disaster actually happens. Thus, this study proposed some community-based and indirect disaster prevention measures. In particular, healthcare is important for victims and necessary for people. It is necessary to strengthen the community network through a daily healthcare network with cooperation among social welfare, medical, and local government agencies at multi levels to maintain good healthcare for local residents after disasters, and to promote health.

Keywords Community disaster resilience • Health promotion • Healthcare • Indirect disaster prevention measure

6.1 Introduction

Disaster is a sudden occurrence that which disrupts the basic fabric and normal functioning of a society (Ideal way and practice of disaster nursing 1998). In particular, natural disasters such as earthquakes, tsunamis, floods, cyclones, droughts significantly impact society, human life, and livelihood (Nakagawa and Shaw 2004). Experiences a disaster can be traumatic, and it can affect people not only physically but also psychologically, socially, and economically (Kako and Ikeda 2009). Disaster damage is caused by the disaster itself, and by loss of private property and health-related public facilities, which affected the local healthcare system. It can result in significant destruction and damage. It is an unpredictable and a sudden phenomenon that is difficult to cope with a domestic or regional level. In addition, victims have difficulties adapting to long-term temporary housing or shelter life. The material damage is of secondary importance, e.g., chronic disease worsens and infectious diseases spread. Therefore, recovery is not just about the rebuilding the infrastructure in an affected area but also concerns rehabilitation of individuals and the process of rebuilding their lives. The importance of long-term mental healthcare has been emphasized, particularly for people who have experienced a disaster (Valentine and Smith 2002). Physical and mental healthcare are needed during the reconstruction/rehabilitation phase after a disaster as well as during daily life. In this way, damage to the victims of a disaster is not only physical but can also have a significant psychological component (Kato et al. 1996; Kessler et al. 1995; Kokai et al. 2004; Toyabe et al. 2007). It is difficult for many victims to maintain their mental and psychological health after a disaster.

Medical assistance teams during the postdisaster acute and sub-acute phases provide medical care mainly at evacuation centers, and mental healthcare teams are dispatched to the affected areas (Nohara 2011). Many victims are rescued and supported by them. However, the survival rate is very low 72 h after a disaster event

(Hayashi 1995). Survivors may feel changes in the mind and body caused by various disaster stresses, which become magnified as time passes. Disaster causes mental stress and psychological stress occurs as a normal reaction to an abnormal situation. The stress response appears in the body, and thoughts, emotions, and actions change over time. Stress reactions in victims of a disaster weaken gradually over time during disaster reconstruction. Adverse effects often appear because of the traumatic experience caused by the disaster. These reactions include hyper arousal, re-experiencing the event, and avoidance. Healing spontaneously within 1 month from such a state is called acute stress disorder (ASD); if the symptoms persist for >1 month, it is called post-traumatic stress disorder (PTSD). In contrast, stress responses during a disaster appear not only in victims but also in the relief referred workers who are referred to as "hidden victims". Relief workers are occasionally guilty and develop a sense of helplessness. Relief activities are performed to fulfill a mission, and these workers forget about themselves while witnessing tragic scenes. However, relief is a panacea, and there is a need to understand that an objective viewing of their behavior is difficult. It is necessary to help other people, the community, and the local society.

Based on these lessons, the euphemisms "kokoro no keaa" literally "care of the spirit", and "kokoro no keaa networking" have become widely used in Japan since the time of the Hanshin-Awaji earthquake in 1995 to exemplify both primary and secondary psychiatric prevention as well as mental health promotion, and mainly to avoid generating the stigma of psychiatric intervention. However, "earthquake-related deaths" in the Hanshin-Awaji earthquake were >10 % of the number of deaths reported (6,400) due to overwork and deterioration from chronic diseases after the earthquake (Hygo Prefecture 2005). Patient information was not shared between healthcare-related facilities and staff because chronic diseases worsened, infectious diseases spread, and disaster stress was experienced in evaluation centers and temporary housing. In addition, many patients' medical records were lost in the areas affected by the disaster. Thus, there is room for improvement in healthcare-related networking. Cuba's medical system is recognized as a good example of a network of daily medical care and disaster medicine. Cuba has successfully implemented community-oriented primary care (COPC), and physicians are held accountable for healthcare outcome measures of the community members they serve. In addition, the medical records in Cuba are organized by families, and health statistics are recorded and reviewed on a regular basis (Presno and Gonzalez 2007). Cuba, where hurricanes occur frequently has focused on disaster medical care as a national policy priority. Support activities using the Cuban medical model of disaster medical care were held in Haiti during the 2010 earthquake (Ono 2012). This system will also help community-coping capabilities when a disaster occurs.

According to the World Health Organization (WHO), health is a state of complete physical, mental, and social well-being and not merely the absence of disease or infirmity. In addition, "healthcare networking" is the smooth exchange of information and services among individuals, groups, and medical organizations that aims to improve, maintain, and promote health. WHO has proposed "Health Promotion", as the art and science of helping people discover the synergy between their core passions and optimal health, and enhancing their motivation to strive for optimal health and supporting them in changing their lifestyle to move toward a state of optimal health (O'Donnell 2009). In Japan, the Ministry of Health, Labor and Welfare (MHLW) proposed "Health Japan 21". The proposed goals of "Health Japan 21" are expected to be expanded or reviewed as occasional demands. In addition, "Health Japan 21" has directed specific plans for health promotion under the heading "Promotion of health in communities" which is to be formulated in a manner best suitable for a particular area (local plans). The plan is being implemented by the cooperation of residents and various community health organizations (Sakurai 2003). Health promotion activity highlights cooperation between individuals, families, communities, and healthcare-related organizations.

Previous studies suggest that primary prevention is the most effective means of reducing earthquake casualties (Dukin 1992). Lessons were also learned, such as the importance of evacuation measures, daily training, and the role of stronger bonds among communities. These bonds need to be utilized for the development of the city and for reconstruction after a disaster. However, it is difficult to maintain awareness of disaster prevention because disaster memories of victims are only forgotten gradually (Atsumi 2007). Japan is expected to have significant earthquakes in the near future, and a "super-aging" society essentially makes the society more vulnerable (Kato 2008). Therefore, preparedness for future mega disasters is essential. Atsumi has proposed "Preventing Disaster Without Saying Disaster Prevention" (Atsumi 2007). Disaster prevention measures will maintain awareness for disaster preparedness through activities such as festivals and healthcare, and these are most important for activating areas of organizational activity (Ojima 2008). Strengthening the healthcare network as an indirect disaster preventive measure will play an important role when a disaster actually happens.

This study considers building a disaster resilient community by utilizing systems and concepts of indirect disaster prevention measures such as Community Health Work (CHW), Community Social Work (CSW), and integrative medicine. This study will document existing practices to strengthen the community network by building a health network improving local disaster risk reduction ability by strengthening prehospital emergency care at the local level utilizing systems and concepts such as first-aid, first responders, Community Emergency Response Teams (CERT), and "Jishu-bosai-soshiki" as coping strategies for the local community against disasters and emergencies.

6.2 Disaster Impact on Physical and Psychological Stress to Victims

Natural disasters elicit physical stress, psychological stress, physical trauma, and economic loss to victims and impact on their health and lives. The physical and psychological stresses on victims differ depending on the types of disaster (Table 6.1). The types of injuries, illnesses, and psychopathology are also different. In addition, these impacts on victims differ not only disaster types but also by disaster timing, life location postdisaster, and age of the victims.

Table 6.1 Disaster impact on the physical and psychological stress

Natural disaster types	Cause of casualties	Physical trauma and disease
Earthquake	Physical trauma by structure collapse	Crushed wound, laceration, fracture, incised wound, cut wound, spinal cord injury, thoracic and abdominal trauma, dust inhalation, burns, crush syndrome, compartment syndrome, bleeding, etc.
Tsunami	Drowning/collision with floating object	Choking, accidental swallowing, pneumonia, head trauma, spinal cord injury, bleeding, bruises, abrasions, incised wound
Flood/typhoon	Drowning/collision with flotsam and falling objects	Choking, pneumonia, head trauma, spinal cord injury, bleeding, bruises, abrasions, incised wound
Volcanic eruption	Contact with the pyroclastic flow/inhalation of hot gas/ inhalation of volcanic ash	Burn, inhalation burn, hydrogen sulfide gas poisoning, respiration disorder
Tornado	Collision/fall	Head trauma, chest trauma, soft tissue injury, bruise, fracture
Prolonged type	Damage and impact is prolonged for taking a long time to rescue the patient and confirmation of the disaster site	Heat stroke, hypothermia, hypertension, crush syndrome, mental disorder, pulmonary thromboembolism, infectious diseases

6.2.1 *Effect of Time*

Physical trauma frequently occurs during the postdisaster acute phase (Table 6.2). In constant, mid-term disasters, worsening of a chronic disease due to insufficient management and infectious diseases due to the decline of immune function also occur. Worsening of mental symptoms due to stress is a problem. For example, in case of an earthquake, bruises, fractures, physically traumatic incised wounds, lacerations, crushing injuries, and head and abdominal trauma due to a building collapse occur mostly early in a disaster, whereas problems such as hypertension; dyslipidosis; diabetes; worsening of chronic diseases, such as liver disease and alcoholism, due to stress and mental illness, such as post-traumatic stress disorder, often occur mid-disaster.

6.2.2 *Effect of Location*

Effects on health differ depending on whether the postdisaster occurs when victims are in temporary housing, a shelter, or their home. Health problems such as insomnia, abnormal bowel movements, loss of appetite, headache, and infection occur in groups living in shelters. Although victims are kept in temporary housing, physical and psychological symptoms appear frequently because of life in an unfamiliar place and anxiety about the future.

Table 6.2 Modulation of physical and psychology in post-disaster phase

Phase	Physical	Thinking	Emotion	Action	Characteristic symptoms
Acute (a few minutes to few days)	Heart rate and breathing increase, blood pressure elevate, sweating/trembling, dizziness, syncope	Hard equations and rational thinking, narrow thinking, poor concentration, hypomnesia, impaired judgment	Dazed, fear, anxiety, sadness, anger	Irritability, restlessness inflexibility, reproachful, communicative competence decline	Struggle and escape reaction
Reaction period (1–6 weeks)	Headache, lumbago, accumulated fatigue, nightmares, sleep disorders	Begin to understand the situation of their own	Sadness, painful and fear recall, depression feel, sense of loss, guilt, mood elevation	Fear of return to the affected site Alcohol intake increase	Suppressed feelings gush
Repair period (1–6 months)	As well as reaction period Intensity of symptoms gradually reduce	Autonomous thinking gradually	Sadness, loneliness, anxiety	Avoid approaching of affected area	Thinking daily life and future, disaster memory and painful emotion recall
Reconstruction period (after 6 months)	Victims will be able to accept own experience without causing a stress response even looking back the disaster events, prepare for other stresses. However, there will be differences in the recovery process by individual victims				

6.2.3 Effect of Age

Children and adolescents develop emotional scars as the result of the disaster experience. Caution is needed because growth of their psyche can be inhibited. Adults are expected to act as agents of reconstruction in affected areas after a disaster. In addition, adults are forced to take care of children and the elderly, which causes stress due to responsibilities to the family and society. The elderly often require need healthcare because they often suffer from various chronic diseases. Health problems often occur as victims cannot adjust to changes in the environment because of life in a shelter or prolonged living in temporary housing.

6.3 Social Backgrounds in Japan

6.3.1 "Super-Aging" Society

Japan is experiencing population aging that is unprecedented in the world. Aging is not only an immediate personal issue but also a salient factor in crucial public policies, such as pensions, health, and long-term care. The low birth rate problem and the diversification of values, and "mutual-help" capabilities dilution of solidarity in the region is reduced, the lives of residents directly related to crime and disaster prevention, environmental protection, such as parenting support has occurred. Under this social circumstance, the East Japan Earthquake and Tsunami (EJET) has highlighted current and emerging issues of a "super-aging" society, especially the need for community-based support systems. In 2000, Japan implemented a universal social long-term care insurance system, under the slogan, "from care by family to care by society" (Campbell and Ikegami 2000). This historic policy has made a variety of home, community-based, and institutional services, a universal entitlement for every Japanese person aged 65+ years based strictly on physical and mental status, regardless of family availability and economic status. It is critical to develop and strengthen community-based support systems, especially for those with limited physical and cognitive function. Labor participation among older adults is essential for sustaining the Japanese society. Problems such as depopulation and aging are faced by the community in Japan. Loss of a sense of community and regional sustainability of local communities are a concern. Assistance by regional authorities will become increasingly important in the future (Yamamoto et al. 2010).

6.3.2 Depopulation

As labor participation increased among women, they delayed marriages or stayed single. Even among the married, fertility rates declined because of the lack of societal support for working women to have children as well as increased financial burdens of raising children. Since the late 1990s, the total fertility rate has been consistently low (1.37 in 2009), much below the replacement level for a population. The total population of Japan in 2004 peaked at 128 million and projects to shrink to 75 % of its peak size by 2050. The results are reflected in the top-heavy population pyramid. In 2030, one person aged 65+ years will be supported by two working-age persons compared with 11.2 and 2.9 persons in 1960 and 2009, respectively (National Institute of Population and Social Security Research 2010a; Statistics Bureau 2003). Japan's population is aging and declining in size simultaneously. Remarkably, population aging is no longer limited to rural areas that suffer from outmigration of younger people; it is an urban phenomenon. Japan precedes other countries in embracing population aging as an urban issue (Muramatsu and Akiyama 2011).

According to the 2010 Population Census, which serves as the base year of these projections, the total population of Japan in that year was 128.06 million (total population including non-Japanese residents). Based on the results of the medium-fertility projection, Japan is expected to enter a long period of population decline. The population is expected to decrease to around 116.62 million by 2030, fall below 100 to 99.13 million in 2048, and drop to 86.74 million by 2060. Based on the results of the high-fertility projection, the total population is expected to drop below the 100 million mark by 2054 to 99.62 million, and to decrease further to 94.60 million in 2060. Conversely, based on the low-fertility projection, the total population is expected to fall below 100 million in 2044 and to decline to 79.97 million by 2060. The range (difference between high and low variant) of the projected total population size by 2060 is thus equal to 14.63 million (National Institute of Population and Social Security Research 2012).

6.3.3 Reduction of Number of Physicians and Medical Facilities

The resulting shortage of doctors in Japan has inevitably led to deterioration in the quality of care (Pronovost 2002; Kahn 2007), and has recently become a serious social problem. The per-capita number of medical doctors in Japan is low compared with those in other developed countries. Japan ranks 59th among WHO 193 member states in terms of number of medical doctors per 1,000 persons (World Health Statistics 2009). Healthcare utilization in Japan is particularly high: the number of consultations per capita is higher in Japan than in any other OECD country (Organization for Economic Co-operation and Development 2009), and the rates of hospital utilization are high as well. These trends have made the shortage of physicians quite obvious. Regarding eventual surplus/shortage of other kinds of health workforce, especially nurses. Some studies have reported a shortage of nurses today (Buchan and Aiken 2008). However, just as for doctors, demand for them will decrease with a declining population in long term.

Through advances in medical treatment technology, the best medical care available can be given, yet at the same time this can lengthen the period of care. In addition, with the progressing trend toward nuclear families and women entering the work force, caring for the elderly at home has become difficult for some households. Concomitantly, there is a shortage of facilities such as nursing homes to care for the aged. This has led to the aged, who primarily require more nursing care than medical treatment, being cared for at hospitals for long periods of time rather than at nursing care facilities, thus accelerating the increase in medical expenditures for the elderly (Health Care 2012).

6.4 Health Promotion and Healthcare Networking

6.4.1 "Health Promotion" and "Health Japan 21"

According to the WHO, "Health Promotion (HP)" is the process of enabling people to increase control over, and to improve their health. The definition of HP provided in the 1986 Ottawa Charter is the most widely accepted definition of health promotion. HP is the provision of information and/or education to individuals, families, and communities that-encourage family unity, community commitment, and traditional spirituality, which make positive contributions to their health status. HP is also the promotion of healthy ideas and concepts to motivate individuals to adopt healthy behaviors. HP represents a comprehensive social and political process, it not only embraces actions directed at strengthening the skills and capabilities of individuals, but also action directed towards changing social, environmental and economic conditions so as to alleviate their impact on public and individual health. HP is the process of enabling people to increase control over the determinants of health and thereby improve their health. Participation is essential to sustain HP action. These strategies are supported by five priority action areas as outlined in the Charter for HP: (1) build healthy public policy (2) create supportive environments for health (3) strengthen community action for health (4) develop personal skills, and (5) re-orient health services. As an activity way for the realization of goals, these active cooperation will develop concrete HP. In so doing, the Charter became the founding document for HP. The third health promotion action area, "strengthen community actions", was explained as follows: HP works through concrete and effective community action in setting priorities, making decisions, planning strategies and implementing them to achieve better health. At the heart of this process is the empowerment of communities-their ownership and control of their own endeavors and destinies (World Health Organization 1986).

In Japan, "Health Japan 21" was developed from the background with a rapid aging of society as a result of the declining birth rate, Japan faces with the fear that life style-related diseases may increase healthcare costs and the burden of nursing care of the elderly people in the twenty-first century. Since it is expected that the aging society would further increase morbidity and the burden of nursing care, and also that economic growth may not continue, it will become important to reduce social burdens related to disease prevention/treatment, and nursing care. Under this circumstance, MHLW established the "National Health Promotion in the 21st Century (Health Japan 21)" in March 2000, in order to create a vital society where

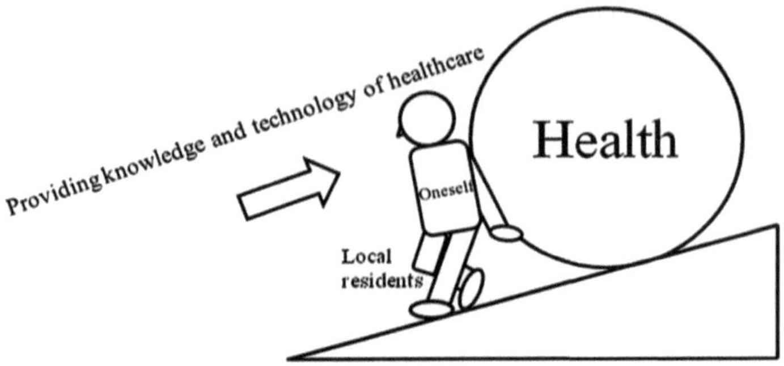

Fig. 6.1 It is difficult to maintain health by oneself

all nationals can live healthy and fulfilling lives (Shibaike et al. 2002). The missions of "Health Japan 21" are to reduce late middle-age deaths, to extend healthy life expectancy and to improve quality of life (QOL). These missions constitute the concept of HP, and objectives with target values to be achieved have been established for several focus areas life styles and specified diseases. Conventionally, various surveys have proposed their own indicators, but the establishment of "Health Japan 21" clarified goals which can be monitored in a systematic way. "Health Japan 21" also encourages the involvement of the society as a whole in the program, making suggestions to health-related organizations etc., in order to improve each individual's awareness of HP and facilitate each to address the issue independently. "Health Japan 21" continues as second phase from since 2012. The four basic policies of "Health Japan 21"are as follows: (1) the importance of primary prevention; (2) creation of a supportive environment for the enhancement of health; (3) goal setting and assessment; and (4) the promotion of effective, well-coordinated activities by the various implementing bodies. The goals for promoting the health of participants should be individual and specific, based on the participant's health status and view of good health, also with reference to the overall goals of the project. The proposed goals are expected to be expanded or reviewed as occasion demands. In addition, under the heading "Promotion of health in communities", "Health Japan 21" has directed that specific plans for the promotion of health be formulated in the manner best suited to the actual situation of that particular area (local plans) by enlisting the cooperation of residents and various community health organizations, in order to effectively promote the plan (Sakurai 2003).

It is important to maintain and promote health. However, it's very difficult to maintain health by only oneself (Fig. 6.1). Therefore, moderate exercise and strength training have benefits such as decreasing the risk of fracture in the elderly and providing good mind values and antiaging effects (Gregg 2003; Lee et al. 2009; Beaglehole et al. 2011). They may also reduce medical costs and health disparities (Wen et al. 2011). Regular daily physical activity should be recommended to elderly (Korpelainen et al. 2010). A high level of healthcare awareness leads to self-care and self-help. The importance of social relationships for health has been well established. At the individual level, social support and networks have been linked to the

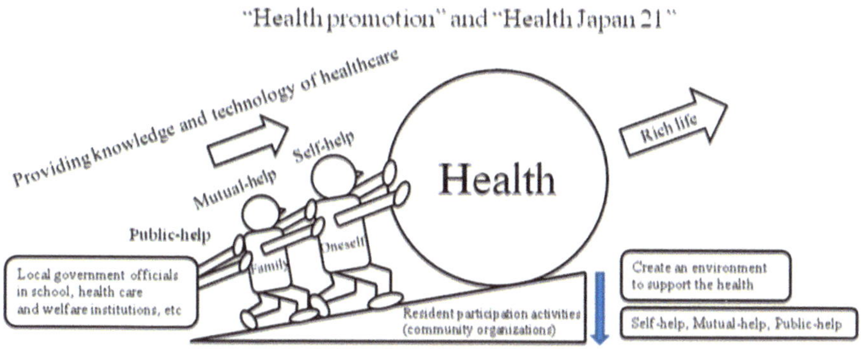

Fig. 6.2 Health is supported by oneself, family, and society

maintenance of health and coping with stress and illness (Berkman and Syme 1979; Cohen and Wills 1985; Uchino et al. 1996). Supporting individuals, families, communities, and the society are important for promoting health and building a rich community (Fig. 6.2) (Toriuchi 1987; Yoshida and Hujiuchi 1995: Reorganization).

6.4.2 Healthcare Networking in Japan

Japan obtained universal public pension and health insurance coverage in 1961 (National Institute of Population and Social Security Research 2011). Long-term care for older adults has been a priority for the past two decades. However, a chronic shortage of rural physicians continues, and elderly adults have not benefited enough from modern western medicine. Public health nurses must support the health problems in rural regions. Wakatsuki has been performed at Saku General Hospital since 1959 for healthcare networking in all of Yachiho village (Nagano prefecture) during this period. This demonstrated that lowering healthcare costs throughout the region could prevent possible diseases and was the first model of the resident's screening system institutionalized in Japan that was ahead of the rest of the world (Wakatsuki 1971).

Kuroiwa worked on community health in Yamato-Machi (currently Minamiuonuma City), Niigata Prefecture in the 1980s. This concept called "Health Yamotopia" includes integrative medicine and effective healthcare networking. Despite long-term care insurance that began 20 years ago, he was actively engaged in home care. At that time, he established a department of Japanese and Chinese medical treatment at a general hospital. He also developed an herb garden with local plants to incorporate acupuncture and traditional Chinese medicine into the hospital. "Health Yamotopia" worked throughout the town center, and this hospital conducted the following programs: (1) overall medical care and health counseling, (2) health seminars, (3) experiential learning (participation in organic farming), (4) practice of herbal medicine, (5) spiritual practice of eating at the Hakkaisan shrine, (6) body operation method for a healthy "dojo," and (7) climbing Hakkaisan. The length of these programs was 7 days. They ate edible wild plants as local cuisine at the dormitory and healed the mind with herb baths in the Urasa hot spring. It is interesting that this project was also intended as salary for this man in Tokyo as well as local residents (Kuroiwa 1987). The approach to prevention and HP occupies an important position in the medical community. Today, healthcare activities with the goal of preventing and promoting health have been performed in various regions of Japan while taking advantage of regional characteristics (Tsuruoka 2012). In this way, professional organizations and healthcare providers such as hospitals, clinics, and health centers, Social Welfare Councils, pharmacies, physicians, nurses, and social welfare workers will cooperate and engage effectively as the daily healthcare system in some regions of Japan.

These cases in Japan represent good healthcare and effective healthcare networking. It is difficult to respond to local resident healthcare needs using only individual or healthcare-related organizations because strengthening daily healthcare networking will play an important role in the community when a disaster occurs.

6.5 Systems and Concepts as Community Based Disaster Prevention and Health

6.5.1 Community Based on Disaster Resilience

During the United Nations International Decade for Natural Disaster Reduction (UN IDNDR 1990–1999), a paradigm shift was observed from postdisaster relief and rescue to predisaster mitigation and preparedness efforts (Nakagawa and Shaw 2004). Previous research suggests that primary prevention is the most effective means of reducing earthquake casualties (Dukin and Thiel 1992). As more research on development has been conducted in various fields in recent years, the approach to disaster mitigation is becoming more and more community-based (Blaikie et al. 1994; Twigg and Bhatt 1989; Mileti 2001; Shaw and Okazaki 2003), and much more effort has been put into incorporating disaster management aspects into the holistic development of communities. In the Hanshin-Awaji Earthquake of 1995, more than 6,000 precious lives were killed. Of the approximately 35,000 people were required to be rescued by the damage of the earthquake, 27,000 people in about 80 % were rescued by community neighbors and local residents (Kawata 1997). There are elderly and victims to need to maintain healthcare in affected area after disaster. In order to promote the reconstruction and development from disaster, and minimize the damage of the next catastrophe, it is necessary to focus on "local communities" when engaging in regional construction in the disaster-affected areas. This should be based on a concept of "disaster reduction", rather than envisioning communities that can be completely prevention to large-scale natural disasters.

EJET highlighted the importance of community-based support systems and emergency preparedness (Muramatsu and Akiyama 2011). After EJET, "kizuna" as meaning bonds or ties between people was often used in Japan. The word as bonds between people in Japan proved strong and resilient. A new awareness of the ties in society seemed to be emerging. The efforts in the face of disaster showed the importance of human connectedness. Furthermore, Japan is a "super-aging" society and because further aging is expected in the near future. Under this circumstance, despite assistance by regional authorities, help will become increasingly important in the future, as the decline in the regional disaster prevention force, which was traditionally held with the assistance of the community, has become a concern (Yamamoto 2010). Therefore, a network of cooperation among social welfare and medical and local government agencies should be further explored to build disaster-resistant communities.

6.5.2 *Sustainable Awareness of Disaster Prevention*

Disaster memories will be forgotten along with the date of the disaster, which may occur in approximately 10 years (Atsumi 2007). Disaster prevention programs in general consist of conducting fire drills, organizing local groups specified in disaster prevention activities, and establishing database systems of disaster experts. It must be effective in an actual disaster. Only such activities often make it difficult for local residents to keep getting interested in "Disaster Prevention" during the peacetime. Watanabe pointed out that people who have not paid direct attention to disaster prevention during the peacetime could have the potential power of disaster management and show their capabilities in the actual disaster (Watanabe, Sugiman et al. 1995). He also recommends that we should use an alternative way to cope with danger. Therefore, it is important to maintain the local resident's sense of disaster through healthcare, which is an indirect measure of disaster prevention because most people over a certain age are highly aware of health, exercise, fitness, and health food.

6.5.3 *Indirect Disaster Prevention Measures*

6.5.3.1 Community Health Work (CHW)

Community Health Workers (CHWs) are members of a community who are chosen by community members or organizations to provide basic health and medical care to their community (International Labour Organization 2008). In many developing countries, there are critical shortages of highly educated health professionals (World Health Organization 2006). CHWs are given a limited amount of training, supplies, and support to provide essential primary health care services to the population. Programs involving CHW in China, Brazil and Iran have demonstrated that utilizing such workers can help improve health outcomes for large populations in underserved regions. Brazil undertook a medical plan named the Family Health Program in the 1990s that made use of large numbers of community health agents. Between 1984 and 2000 Iran was able to cut its infant mortality in half and raise immunization rates from 20 to 95 % (Roudi and Bureau 2003). WHO estimates there are over

1.3 million CHWs worldwide (World Health Organization 2011). In India, CHW have been utilized to increase mental health service utilization and decrease stigma associated with mental illness. In New York, CHWs have been deployed across the state to provide care to patients with chronic illnesses like diabetes that require sustained, comprehensive care. They are seen to play an important role in assisting patients with navigating a complex, uncoordinated healthcare system (New York State Health Foundation 2013). With training, monitoring, supervision and support such workers have been shown to be able to achieve outcomes far better than baseline and in some studies, better than physicians (Perez et al. 2008). Important attributes of CHW are to be a member of and chosen by the community they serve. This is crucial because many communities are disengaged from the formal health system. CHW is unable to emigrate because they do not have internationally recognized qualifications (McPake and Mensah 2008). Much remains to be learned about the recruitment, training, functions, incentives, retention and professional development of CHW (African Medical and Research Foundation (AMREF) 2007).

Furthermore, health professional volunteers played an important role after the 1995 earthquake in Hyogo, Japan. From lessons learned, CHW will help victims and assist medical and healthcare professionals when disasters occur.

6.5.3.2 Community Social Work (CSW)

CSW is methods proceed in social welfare activities that focus on community have been proposed in the Berkeley report was published in the UK in 1982. CSW is the contemporary method in community-based social work. In today's community practice, it is important to construct methods which have an inclusive, integrated viewpoint for the independent life support of the individuals in the community. Also, institutional or systematic upgrading becomes necessary to develop such methods (Hanashiro 2002). In recent years, interest in CSW has risen. The project of CSW have started by Osaka Prefecture since 2004 which suggested that CSW should work indirect ways as well as direct ways to help clients, and as 'social care planner' and 'developing networking in the community' rather than merely as clinical social worker. Community social workers (CSWs) assist the people who need assistance in the region by implementing community-based activities with an emphasis on environmental aspects, such as human relations and living space of these people. They have the expertise to manage the relationships between public institutions and new services. CSWs are appointed in junior high schools in each district for the purpose of bridging ties with organizations such as the Social Welfare Council, municipalities, and elementary schools. Hence, they function as "Tsunagi" at both the municipal level and the elementary school level. They are advanced professionals with regional analysis capabilities and the ability to respond to individual issues. In their position, they must fully understand the middle school system, and they should make efforts to connect people at the elementary school and municipal levels (Yasuba 2008). Thus CSW are expected that they should work the direct way and the indirect way. Their main tasks would be both the immediate well-being of their clients and aiming to strengthen coping networks in the community in order to ensure that people at risk were better cared for, which has been until now the task of community worker.

CSW will promote community-based welfare, help victims, assist welfare-related facilities and support social welfare workers when disasters occur.

6.5.3.3 Integrative Medicine

Integrative medicine is based on understanding the whole person, not just the physical symptoms. Integrative medicine is more than just adding up conventional and alternative therapies. According to the WHO, inhabitants of 80 % use traditional medical care in the African area, and Beninese as a high country of the use rate (80 %) in particular is mentioned. In addition, recognition that a solution of the problem is difficult only in western medicine spreads out, and complementary and alternative medicine (CAM) is developed. And also in the developed country, the necessity of integrative medicine that fused in them more is advocated by a change of the disease structure, the increase of the medical cost, a change of the recognition of the patient (Nagahori 2011). Integrative medicine is a phrase coined by Andrew Weil, he goes on to clarify that integrative medicine as a healing-oriented medicine that takes account of the whole person (body, mind, and spirit), including all aspects of lifestyle. It emphasizes the therapeutic relationship and makes use of all appropriate therapies, both conventional and alternative. Integrative medicine envisions a health care system that focuses on efficient, evidence-based prevention, wellness, and patient-centered care that is personalized, predictive, preventive and participatory (Eisenberg et al. 1993; Eisenberg 1998). Integrative medicine also encourages patients, doctors, and complementary practitioners to work as a team for the best possible outcome (Peters and Woodham 2002).

Cuba is one clear example in which COPC as Cuban family medicine has been successfully implemented, and physicians are held accountable for health care outcome measures of the community members they serve (Presno and Gonzalez 2007). Cuba's COPC and CAM are well developed within the Cuban medical system. The Cuban health system is of special interest because of its developed-country health outcomes despite its developing-country economy (Hood 2000; Dresang et al. 2005). The current system of family medicine based in neighborhood consultorios was established in 1984. Family physicians, paired with nurses, serve approximately 600 patients or 150 families in a defined geographic area surrounding their "consultorio". Consisting of interdisciplinary teams, policlinicos offer specialty care in a variety of areas, usually including pediatrics, internal medicine, nursing, social work, dentistry, and physical therapy, and sometimes including cardiology, pulmonology, ophthalmology, neurology, endocrinology, dermatology, and psychiatry. Much of what is called CAM.

After EJET, acupuncture/massage therapy, which has both mental and physical soothing effects, may be a therapeutic approach that can be effectively used in combination with western medical practices. In Japan, recently the academic, senators, and citizen's organization started to promote integrative medicine (Atsumi 2008). The aging society that recent medical development drought complicates this problem. So much expectation is sent to alternative medicine (Shimada 2009). Acupuncture, yoga, music therapy, and aromatherapy have played a major role after the EJET (Takayama 2012). Based on this observation, integrative medicine concepts will be needed in the future.

6.5.4 First-Aid

First-aid is the provision of initial care for an illness or injury. It is usually performed by non-expert, but trained personnel to a sick or injured person until definitive medical treatment can be accessed. It generally consists of a series of simple and in some cases, potentially life-saving techniques that an individual can be trained to perform with minimal equipment. The key aims of first-aid can be summarized in three key points; (1) preserves life (cardio pulmonary resuscitation (CPR), (2) prevent further harm, (3) promote recovery. First-aid training also involves the prevention of initial injury and responder safety, and the treatment phases. The first aider is also likely to be trained in dealing with injuries such as cuts, grazes or bone fracture. They may be able to deal with the situation in its entirety, or may be required to maintain the condition of something like a broken bone, until the next stage of definitive care (usually an ambulance) arrives. Training is generally provided by attending a course, typically leading to certification. Due to regular changes in procedures and protocols, based on updated clinical knowledge, and to maintain skill, attendance at regular refresher courses or re-certification is often necessary. Trainees may also learn the basics of automated external defibrillation (AED). First-aid training is usually available through fire department, the Red Cross, and private facilities in Japan. General training course is free or require a small fee. Many courses have training times of approximately 3 h. People who receive first-aid training contribute to improving the survival rate as bystanders (Kano et al. 2005).

Large-scale disasters such as earthquakes cause plenty of injured people and urgent medical aids increase rapidly. Basically, medical resources such as doctors, nurses, and Emergency Medical Technicians (EMTs) are limited. Therefore, local inhabitants' participation as initial responders is need under large-scale disasters. The first-aid lesson might provide effectively person the participation awareness under emergency situation comparing to the large-scale disaster trainings such as earthquake trainings because of the first-aid class for the situations such as injuries or cardio pulmonary arrest (CPA), which needs first-aid occurs more than large-scale disasters. It is also recognized that members of the lay public are often the actual first responders in many disaster events (Auf 2003). In the 1990 earthquake in Luzon, Philippines, most of the crucial life-saving rescue work was carried out by lay members of the local community (Roces 1992). In California, the government expects residents to be self-sufficient for the first 72 h after an earthquake. Thus, several researchers have recommended that members of the lay public receive first-aid training in order to develop a community's capability to provide basic medical assistance during the immediate postdisaster phase until professional assistance

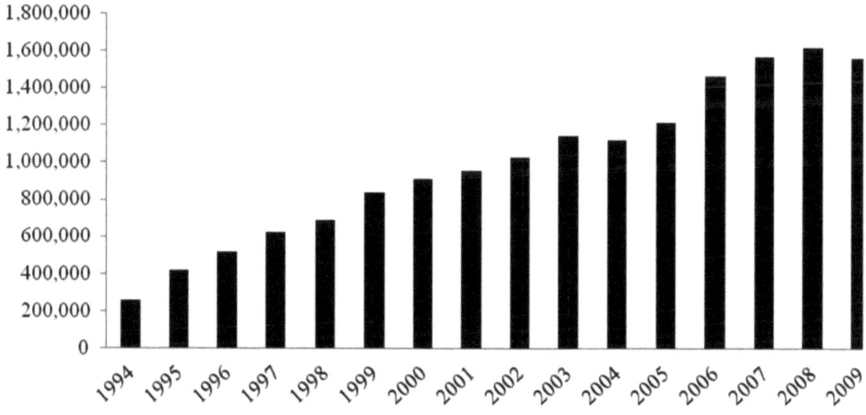

Fig. 6.3 Transition of participants number of first-aid training in Japan

becomes available (Angus 1992, 1993; Crippen et al. 2011). Lay members trained first-aid of the public can potentially contribute to a postdisaster medical response (Kano et al. 2005). Dr. Loc Nguyen et al. examined the prevalence of first-aid training in a community-based sample in Los Angeles, identifying demographic factors that predict self-perceived first-aid ability. They discovered that, while first-aid training was the most influential factor predicting high self-perceived first-aid ability, ethnicity, age, language preference and income were also predictive variables (Kawachi and Subramanian 2006). First-aid has been recommended by the Ministry of Education, Sports, Culture, Science, and Technology (MEXT) and the Fire and Disaster Management Agency (FDMA) in Japan. First-aid training is also in high demand from the citizens (FDMA 2011 Fig. 6.3), as it may lead to self-help, mutual-help, and coping with emergencies.

6.5.5 *Coping Capacity of the Local Community Against Disaster and Emergency Response*

6.5.5.1 First Responder

"A certified first responder" is a person who has completed a course and received certification in providing prehospital care for medical emergencies. They have more skill than someone who is trained in basic first-aid but they are not a substitute for advanced medical care rendered by EMTs, emergency physicians, nurses, or paramedics. First responder support EMTs and Paramedics, provide basic first-aid, CPR, AED use, spinal immobilization, oxygen, and assist in emergency childbirth. The certified first responder is a generic term referring to the first medically trained responder to arrive on scene (police, fire, EMS). Most police officers and all professional firefighters in the US and Canada, and many other countries, are certified first responders.

Community First Responder (CFR)

CFR are a person available to be dispatched by an ambulance control centre to attend medical emergencies in their local area. They can be members of the public, who have received basic training in life saving interventions such as defibrillation, off duty ambulance staff, or professionals from a non-medical discipline who may be nearby or attending emergencies, such as firefighters or security guards. CFR are found in the emergency healthcare systems of the United Kingdom, the USA, and Australia (Jackson 2004). In general CFR are there to provide assistance to those with a medical emergency, and most importantly to start and maintain the chain of survival in CPA patients until a fully equipped ambulance arrives. The majorities of responders are volunteers and take no payment and use their own cars with no mileage recompense.

6.5.5.2 Community Emergency Response Teams (CERT)

CERT programs began in Los Angeles in 1993 and there are teams in a number of states. The CERT members (CERTs) are to be trained by Federal Emergency Management Agency (FEMA) directly or through state programs so that the teams can support local first responders (Willian and Waugh 2003). One example of a community-based disaster preparedness program that trains members of the public in first-aid is CERT, which originated in Los Angeles, California, and is currently implemented in 48 US states and 6 foreign countries. This program focuses on the assessment and treatment of precisely those injuries that most commonly occur in natural disasters, namely, broken limbs, puncture wounds, lacerations or burns (Simpson 2000). Following a major, first responders who fire and medical services will not be able to meet the demand for these services. In the United States, CERT can refer to an organization of volunteer emergency workers who have received specific training in basic disaster response skills, and who agree to supplement existing emergency responders in the event of a major disaster. A local government agency, often a fire department or emergency management agency, agrees to sponsor CERT within its jurisdiction. When not responding to disasters, CERTs may raise funds for emergency response equipment in their community; provide first-aid, crowd control or other services at community events. A team may self-activate when their own neighborhood is affected by disaster. An effort is made to report their response status to the sponsoring agency. A self-activated team will size-up the loss in their neighborhood and begin performing the skills. They have learned to minimize further loss of life, property, and environment, will continue to respond safely until redirected or relieved by the sponsoring agency or professional responders on-scene. In the short term, CERTs perform data gathering, especially to locate mass-casualties requiring professional response, or situations requiring professional rescues, simple fire-fighting tasks, light search and rescue, damage evaluation of structures, triage and first-aid. In the longer term, CERTs may assist in the evacuation of residents, or assist with setting up a neighborhood shelter. While responding, CERTs are temporary volunteer government workers.

FEMA recommends that the standard, ten-person team be comprised as follows: CERT Team Leader, Safety Officer, Fire Suppression Team (2 people) suppressing small fires, Search and Rescue Team (2), Medical Triage Team (2) and Medical Treatment Team (2). Because every CERT member in a community receives the same core instruction, any team member has the training necessary to assume any of these roles. This is important during a disaster response because not all members of a regular team may be available to respond. Local residents will be the first responders. However, the process of organizing local residents must take place before, during, and after such catastrophic events occur (Berke 1993). Successfully linking local organizations, citizens, and leaders provides a strong network and a method for local citizens and groups to become actively involved in local preparedness and response efforts. To be most effective, this process of capacity building must take place before disasters occur, and continue during and after such catastrophic events. CERT skills are useful in disaster and everyday life events. CERT training includes disaster preparedness, disaster fire suppression, basic medical operations, and light search and rescue operations. Local volunteers and community level action are essential to effective natural disaster preparation and response (Brennan 2005). CFRs and CERT will take effect at the time of a disaster as well as emergency responses at prehospital. FDMA is preparing to adopt some regions in Japan, as the need has been advocated. CFRs and CERT provide not only prehospital care but also the disaster-coping capacity of the local community.

6.5.5.3 Jishu-bosai-soshiki

Jishu-bosai-soshiki, or Jishubo for short, literally meaning "autonomous organization for disaster reduction" is a neighborhood association for disaster preparedness and rescue activity at the community level in Japan. Jishubo is a form of participation activity particular to Japan (Wachtendorf and Tierney 2001). In the normal times between emergency situations, member households are expected to organize disaster drills, educate residents, patrol their residential area and maintain rescue tools. During emergencies, Jishubo members guide refugees to a shelter, rescue residents, provide the initial first-aid and supply food and water. The Jishubo system has become in practice a government campaigned community-based disaster management organization and as such, it has been achieving a certain level of effect on disaster reduction as a whole. The increase in the number of Jishubo organizations implies a favorable growth of interest in Jishubo among Japanese households, and there has been a steady increase in the number of participating households, especially after the Kobe earthquake in 1995 (Bajek 2008). However, the disadvantage is that most of the participants in Jishubo are elderly people who may be participating in Jishubo for reasons not entirely related to disaster preparation/mitigation and thus do not truly want to participate in the social scheme of Jishubo. New social approaches that can be implemented simultaneously are needed to supplement this weakness of such government-initiative participation. As Shaw and Goda pointed out (Shaw and Goda 2004), the Kobe earthquake caused another social change in terms of the renaissance of volunteerism and naissance of non-profit organizations (NPO) as new conduits of social activities. These organizations are not based on Chonaikai, but are an "association" of motivated members, and it may be effective for future Jishubo to cooperate with such organizations. The quality of Jishubo's activities, including the contents of the activities or connections with other social organizations, will thus be given more of a role of an indicator of disaster resilience rather than merely an organizing rate (Gough 2000). Sense of solidarity established in the region will become a valuable asset not only useful as encounter a disaster but also in order to build more healthy and more comfortable rich life.

6.5.6 Evaluation of Healthcare and Community Disaster Coping Capacities

6.5.6.1 Noriko Yoshida Method (NyMthod)

NyMthod is a holistic health care program developed by Noriko Yoshida. The purposes of the program are to evaluate body, mind, spirit, social health, and self-care ability as well as to structure a community-based holistic model and propose an integral map for holistic health promotion in the community. In addition, the program satisfies individual needs and collective individual health education (method of collective individual approach) that will lead to building regions that support their own health. The program is a course consisting of seven programs to assess the increase in an individual's holistic health and to obtain the capability for self-help, mutual-help, and regional potential. Improvement in local resident's healthcare can be evaluated using this method. In addition, evaluation and improved self-help, mutual-help, and regional potential may lead to increasing coping capacity of local communities against disasters and emergencies.

6.5.6.2 Diagnosis of Coping Capacity of the Local Community Against Disaster Developed by Cabinet Office and FDMA

(1) Diagnosing the coping capacity of a local community against landslides (Cabinet Office). (2) Diagnosing the coping capacity of a local community against flood damage (Cabinet Office). After a regional disaster, the prevention leader answers questions in a questionnaire format. They develop a comprehensive evaluation of the coping capacity of the local community against disaster and the strengths and weaknesses of each region using graphs and comments (Japan Cabinet office 2003). (3) Evaluation guideline for the coping capacity of local governments against disasters and crisis management developed by FDMA. The evaluation guideline intends for local governments to take on the responsibility for their own disaster management system. Each local government answered approximately 800 questions (Yes

or No, or four select). The guideline evaluated disasters response by representing the results with various graphs.

6.5.6.3 Diagnosis of Regional Disaster Prevention Ability Developed by Matuda et al.

This diagnostic method is intended to support disaster prevention measures for local residents and to indicate the gap in recognizing the coping capacities of local communities against disaster as a means for NPO prevention experts and local residents. This method is also used to allow two-way communication between experts and local residents. This method emphasizes on creating a "TO-DO list" of their own with cooperation from third-party experts and local residents rather than to present a "TO-DO list" (Matsuda 2005).

6.6 Medical and Healthcare After the East Japan Earthquake and Tsunami (EJET)

A magnitude 9.0 earthquake occurred in the international water of the western Pacific and induced a huge tsunami on 11 March 2011. EJET hit the northeastern part of Japan and caused heavy casualties, heavy property losses (Okada 2011). As of 5 June 2011, the National Police Agency had registered 15,365 deaths, 8,206 missing, and 5,364 people injured across 20 prefectures and 111,044 destroyed and damaged buildings (NPA 2011). Miyagi, Iwate, and Fukushima Prefectures were the worst hit, with 9,184, 4,524, and 1,592 deaths, respectively (NPA 2011), involved with 107, 119, 27 deaths and missing persons of volunteer fire corps (Kahoku Shinpou 2011).

6.6.1 Acute and Sub-acute Medical Activities

After EJET, a total of 128 DMATs, consisting of one to two physicians, one to two nurses and one administrator, came to the assistance of Iwate prefecture for a 9-day period beginning 11 March 2011, engaging in emergency medical care and patient transportation activities and other duties. These teams provided transportation via helicopter to inland areas and established a wide-area distribution centre and staging care unit (a temporary medical facility for emergency medical evaluation outside of affected areas), at Hanamaki Airport further inland from where they provided wide-area transportation and distribution services via aircraft to medical facilities both inside and outside Iwate Prefecture. Taking on the work of medical activities performed by DMATs in the postdisaster acute phase, medical assistance teams (Japanese Red Cross Society, Japan Medical Association, etc.) provided medical care mainly at evacuation centers from the sub-acute through to the chronic phases of operations (Nohara 2011).

6.6.2 Mental and Psychosocial Health Activities

In addition to medical teams, many public health and sanitation teams engaged in activities in the disaster areas. Based on previous experience of natural disasters such as Hanshin-Awaji and Niigata earthquakes, more than 50 mental health care teams have been organized and dispatched to the affected areas, scheduled by MHLW. These included health maintenance activities by public health nurses, mental care from psychiatrists and clinical psychologists, oral care by dentists, as well as support from pharmacists, certified nurses, occupation a therapists and physiotherapists (Nohara 2011). Based on previous experience of natural disasters, team or systems of medical relief and mental healthcare have been established in acute and sub-acute. However, even now, many victims have medical problems that must to be address. Also, it is likely that following a disaster, healthcare relationships will have been disrupted, and medical services will be needed independent of any mental health concerns (Madrid et al. 2008).

6.6.3 Disaster Impact Physical and Psychological Health

Many elderly people require health management for chronic diseases. Therefore, after the EJET, it was difficult for them to adjust to changes in the environment in an evacuation shelter or prolonged living in temporary housing (Yamaguchi et al. 2011). The conditions that should be emphasized in disaster-affected areas include hypertension, deep vein thrombosis, respiratory problems, mental and psychosocial health, and infectious diseases. Many residents in disaster-hit areas lost their family, friends, homes, possessions, and their jobs, and the psychological impact has been immense. After the disaster, reports emerged suggesting cases of PTSD (Nohara 2011). Almost all deaths after EJET were of the elderly. As one reason for death, fatigue of living in a shelter and the physical and mental stress after the tsunami and the earthquake occurred at a high rate. As in most natural disasters, the oldest people were the hardest hit. Older adults with functional disabilities had difficulties escaping the tsunami that killed more than 90 % of the 15,000+ people who lost their lives in the earthquake. Surviving older adults were vulnerable to cold temperatures, influenza, relocation, and mental and physical stress. Many struggled without access to medications and treatments needed to control their chronic conditions, which could result in premature deaths. Persons aged 65+ years account for more than 90 % of the growing number of such "earthquake-related deaths" (524 deaths at 241 hospitals in Miyagi, Fukushima, and Iwate) (Nippon Hoso Kyokai (NHK) 2011). This historic catastrophe occurred in an earthquake-prone "super-aging" society. Its profound implications go far beyond its immediate impact on the most vulnerable.

The importance of long-term mental health care has been emphasized particularly for these people (Valentine and Smith 2002). The lessons from previous disasters such as the great Hanshin-Awaji and the Niigata-Chuetsu earthquakes tell us that the psychological impact of the disaster will require careful, attentive and systematic assistance over the long term (Kuwabara et al. 2008; Suzuki et al. 2011; Shinfuku 2002; Kim and Akiyama 2011) addition, it will be essential to provide sustained assistance to bereaved children and orphans on whom the psychological impact will be especially great, as well as on-the-ground medical activities and technical advice to local support staff. These include healthcare workers, local government officials, fire and police service personnel and teachers who were, despite engaging in assistance activities, disaster victims themselves (Nohara 2011). In this way, disaster damage to victims not only occurs during the postdisaster acute and sub-acute phases but the impact also poses long-term physical and psychosocial problems.

6.6.4 Issues of Medical and Healthcare After Disaster

When disaster first strikes, accidents or emergencies are the most common reasons for people to seek health care. In most cases, the rapid response of local providers of care, disaster medical assistance teams, volunteers, and others helps to assure that accident and emergency patients receive appropriate intervention (Axelrod et al. 1994). The disaster highlighted challenges that Japan is facing especially rebuilding disaster area communities and addressing population aging in urban communities where Japanese traditional qualities are fading. After EJET, however, a health care system, especially one serving the poor, may have been inadequate or overburdened before the disaster, and its ability to respond is often adversely affected by destroyed or damaged buildings and equipment, lack of drugs and other supplies, and increased shortages of healthcare providers (Axelrod et al. 1994). In Japan, patient medical records are not shared between hospitals and clinics in many regions during peace time. Kunii indicated that although surveillance systems and the collection and analysis of information regarding infectious disease, including early warnings and countermeasures, were available, it took more than 1 month to provide relief to some areas. If possible, a system for future catastrophes should be developed within a week (Kunii 2012). On the basis of these causes, the healthcare network may not have been engaged sufficiently and much patient information was lost in affected areas after the EJET, resulting in problems treating patients.

Cuba demonstrates that achieving and maintaining a healthy population with good health results does not necessarily depend on a big budget or richness of a country. Good governmental policies regarding public health and social security are crucial to achieve a good QOL equally distributed to the whole population (Gutiérrez et al. 2003). In Cuba, a growing cadre of family physicians is being trained to work as part of a healthcare team trained to provide health education and preventive services, offer comprehensive medical care, and conduct population based-research. In Cuba, family physicians are required to look at patients in the context of family and community. Medical records are organized by family. Health statistics are recorded and reviewed on a regular basis (Dresang et al. 2005). Cuba which Hurricane occurs frequently has been focused on disaster medical care as a national policy. Support activities by the medical model of disaster medical care in Cuba was held in Haiti in 2010 earthquake (Appelbaum et al. 2006).

By referring to past experiences of national and international large-scale disasters (Emergency and Humanitarian Action 2011), there was an effective response to several health-related challenges pertaining to medical and public health and sanitation. However, there were many issues that exceeded initial expectations or scenarios for a large-scale disaster, and a specific response needs to be made to tsunami disasters. There are many outstanding challenges that cannot be resolved by resorting to a response based on a disaster manual (Nohara 2011). Based on the lessons learned, sharing and preservation of patient information system that utilizes information and communication technology (ICT) is to be constructed by MHLW. A research team has been established. The new system called the "Disaster-resistant cooperation of local healthcare" aims to support patients and share information beyond the boundaries of medical institutions and pharmacies, and many nursing care facilities are digitizing patient information (Mainichi Newspaper 2012). Individual healthcare providers, teams and facilities could have improved postdisaster acute and sub-acute medical care on the basis of lessons learned from previous disasters. However, healthcare networking may not have engaged sufficiently, and some problems related to healthcare and healthcare networking have been pointed out. There will be room for improvement of healthcare networking. Therefore, in order to maintain good healthcare of physical and mental, and reduce "earthquake-related deaths" for local residents after disasters (Kato 2012), it will be necessary to strengthen the community network through daily healthcare.

6.7 Next Disaster

6.7.1 Tokai, Tonankai and Nankai Earthquake

Japan is one of the world's famous countries for natural disaster (Suzuki et al. 2010). Japan is vulnerable to a wide variety of natural disasters such as earthquakes, tsunamis, volcanic eruptions, typhoons, floods, landslides, and avalanches. Of these natural disasters, earthquakes are the most serious and frequently occurring (Sawada 2012). Furthermore, the next great earthquakes are expected to occur in the near future (Kato 2008). Japan is located on a plate boundary, and therefore, earthquakes and tsunami causing extensive damage could occur anywhere in the country. In addition, because most of Japan's population and major resources are concentrated in low-level coastal areas, there is a great potential for damage caused by tsunami. It is necessary to recognize once again that Japan does face serious risks in the form of earthquakes and tsunami. With regard to the measures against the possible "Tokai Earthquake", "Tonankai Earthquake", and "Nankai Earthquake" that could possibly occur in the first half of this century and be accompanied by major tsunami, what have learned from EJET and set a direction for new measures must be incorporated. When faced with a large-scale disaster, the following concepts are all of vital importance: "public-help" offered by the central and local governments; "self-help" taken on by individual citizens and private corporations; and "mutual-help" where by individuals, corporations, and organizations of local communities work together and support each other. Given Japan's aging population and falling birthrate, as well as the advancement of globalization, it is need to also consider those who will assist groups of people such as the elderly and foreigners in times of disaster (Towards Reconstruction, "Hope beyond the Disaster" 2011).

6.7.2 Earthquake in the Nankai Trough

The government unveiled a worst case disaster scenario warning that a magnitude 9 earthquake in the Nankai Trough off Japan's Pacific Coast could kill over 320,000 people, dwarfing last year's quake-tsunami disaster. According to the projections released by the Cabinet Office, up to 320,000 people could be killed in 30 prefectures by tsunami generated by a massive earthquake in the trough which stretches for 750 km from Kanto to Kyushu. The projections were based on a scenario in which a quake strikes at nighttime during the winter with strong winds helping to unleash tsunami up to 34 m high, sweeping many victims away as they slept. Many of the estimated 323,000 victims would be drowned by the tsunami, crushed under falling objects or in fires sparked by the disaster, it said. The highest number of casualties are expected to be in Shizuoka (109,000 projected deaths), followed by Wakayama (35,000), Miyazaki (34,000) and Kochi (25,000). However, the report said the number of deaths could be reduced by 80 % if evacuations begin within 10 min of a tsunami alert. It also called for more evacuation centers on high ground and taller buildings, as well as regular evacuation drills. At the town of Kuroshio in southwestern Kochi Prefecture, the tsunami could reach 34.4 m—the highest level projected under the scenario, the Cabinet Office said. In its previous projection in 2003, the panel gave a worst case scenario in which no areas would be hit by a tsunami of more than 20 m. But the panel has upgraded its predictions in the wake of the 9.0-magnitude earthquake on March 11 last year that sent a tsunami barreling into the northeast, killing some 19,000 people and devastating the coastline. "As long as we live in Japan, we cannot deny the possibility of a huge earthquake and tsunami," Masaharu Nakagawa, state minister for disaster management, told reporters. The report was designed to paint a worst-case scenario and help officials boost their disaster preparedness (Japan Today 2012; Cabinet Office, Government of Japan 2012).

6.8 Future Efforts

Japan is reviewing and reinforcing its emergency preparedness at multiple levels. Rebuilding broken social relationships, reintegrating isolated older adults, and encouraging older adults' labor participation are major challenges in rapidly aging urban communities (Muramatsu and Akiyama 2011). Under "super-aging" society, many older adults want to work: 40 % of workers aged 60+ years wished to continue working for as many years as they can (Cabinet Office 2011). Development of health and medical and nursing and welfare system which focuses on providing comprehensive community care services. For the reconstruction of the affected municipalities, a system of comprehensive community care services will be developed. While the basis is the traditional system of community-centered support, the new system will integrate the provision of health and medical, nursing and welfare, and livelihood support services. Keeping in mind the factors of convenience and disaster prevention, attention will be given to developing integrated facilities and promoting the shared use of facilities such as housing, health and medical facilities, welfare facilities, business offices providing nursing care and welfare, and educational facilities. In particular, with regard to medical services, medical functions should be consolidated and coordinated in view of the shortage of physicians, etc. in the affected municipalities. In this process, efforts should be made to promote the use of home medical care (e.g., Cuban COPC), provide seamless services to address the medical needs of patients, and to facilitate the early recovery and burden reduction of patients. EJET of 2011 highlighted the importance of community-based support systems and emergency preparedness. Such systems should incorporate innovation (e.g., new technology) and collaboration of multiple stakeholders including older residents. Building on existing societal strengths may enhance the effectiveness of new systems to prepare for the future. The earthquake also revealed strengths of traditional social relationships rooted in close family and neighborhood networks. Social networks that encompass multiple generations within families and communities are sources of instrumental support as well as happiness and stress in Japan (Akiyama et al. 1997). Elaborate emergency preparedness plans, which had been in place in local communities and neighborhood mutual-help systems, were effectively activated to facilitate evacuation and coping. Close-knit community-based social networks enhanced provision and receipt of social support. To preserve social support systems in earthquake-stricken communities, local governments have relocated whole communities as intact as possible.

However, the disaster, although certainly tragic, provides opportunities to rebuild communities in innovative ways that accommodate the "super-aging" society and protect against the natural disasters that hit Japan periodically. Building such communities will require close collaboration among local governments, industries, health care providers, academic communities, and residents. In addition, enhancements need to be made to other health-related services, including the use of private companies. Information and communication technologies, etc. will be used be to coordinate health and medical and nursing and welfare services. At the same time, steps need to be taken to facilitate the sharing of clinical information, including patient medical records, for the purpose of future crisis management. In the reconstruction process, residents will newly help each other out in the course of living at evacuation centers and temporary housing, etc. This is expected to encourage residents to further watch over each other and advance social participation. These initiatives in the disaster region will serve as model initiatives for Japan's declining birthrate and aging society in the future. Thus, in non-disaster affected regions, too, it would be desirable to shift to the "model of comprehensive community care services".

This study proposes that local residents learn first-aid skill. Learning first-aid leads to improvement in emergency response capability, as self-help increases during the process of learning first-aid. First-aid training is an opportunity to build healthcare and healthcare networking. Because first-aid training is popular in Japan, it has been recommended by the MEXT and FDMA. For example, Kobe City has promoted efforts to spread first-aid training since 1995 after the Hanshin-Awaji Earthquake. At total of 66 civil societies first-aid organizations teach first-aid to citizens. First-aid may play an important role as a pathway between healthcare staff and facilities during the process of building a healthcare network. Second, disaster prevention and its education training are very important. Furthermore, community ties strengthen through the healthcare systems such as CHW, CSW, and integrative medicine, which may be successful for sustaining community disaster prevention. Interesting healthcare activities may lead to disaster prevention measures. Various healthcare-related study sessions are being conducted in community and health centers because public awareness related to healthcare has increased. Self-help and mutual-help will grow during the process of learning healthcare. Learning about healthcare will be an opportunity for disaster prevention education and building disaster prevention networking.

Third, residents acquire first-aid knowledge and skills, and healthcare will organize CERT and Jishubo, and may be a member of CFR. Thus, emergency coping capabilities and disaster prevention by local communities will improve. Engaging local residents and fire corps in a healthcare network by learning from first-aid workshops strengthens prehospital emergency care at a local level. Enhancing cooperation with local residents and fire corps will assist in building a healthcare network. After building a disaster-resilient community through healthcare networking, it will be necessary to analyze and evaluate the healthcare of local residents and the regional disaster coping capacity. Based on comprehensive support center, healthcare providers will provide healthcare for local residents by depending on individual of their situations in inside and outside regions. Various type healthcares support for them through utilizing CHW, CSW, and integrative medicine. If the local residents could be cared in inside local region, healthcare will need cooperation with among community center, public health center, and healthcare-related institutions in neighboring local area. Even if they could not be cared in inside local region, healthcare will need cooperation with among NPO/NGO in outside area and healthcare-related institutions in remote area. The cooperation is interaction of healthcare providers and healthcare-related information. Hence, in order to build effective healthcare network, it is necessary to cooperate with various healthcare-related providers and institutions in inside and outside area. In addition, community disaster resilient and emergency capabilities of local region will be enhanced and improved through linkage and cooperation with CFR, CERT and Jishubo. Therefore, management of comprehensive support center will be very important in the future (Fig. 6.4).

However, the outcome may not be uniform because different systems depend on the characteristics of each region. Finally, sharing the contents and results of this

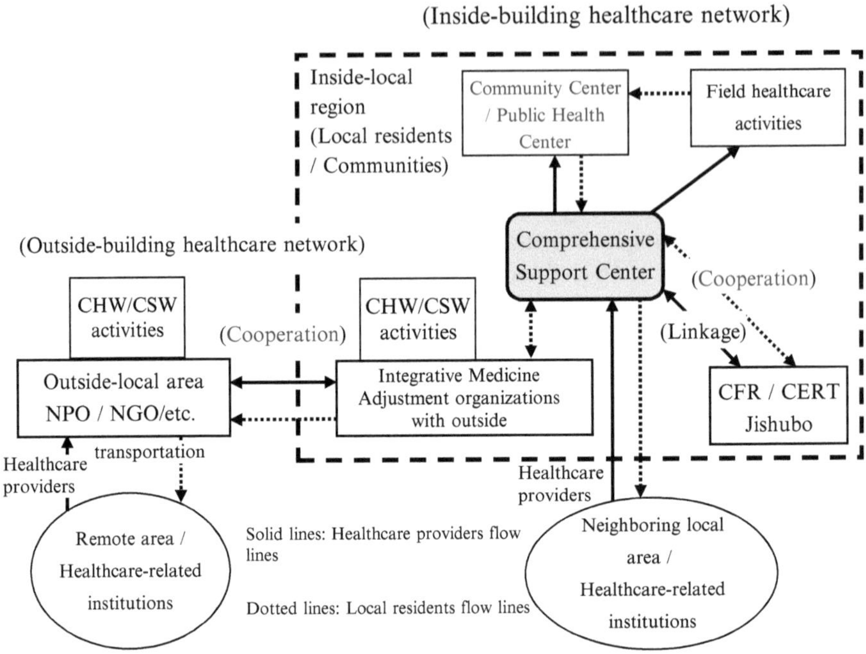

Fig. 6.4 Building healthcare network in inside and outside-local regions

proposal plan will be important. Sharing information with disaster management officials in Asian countries where the risk of earthquake and tsunamis is high will also be necessary. Sustainable cooperation among local communities will be achieved during the process of building a healthcare network. The local prehospital emergency care system will be enhanced by a first responder system, and the emergency response capacity of the community will improve. Healthcare adapted for an ageing society has become a city planning model, both in Japan and abroad.

Japan faces major challenges stemming from simultaneous population aging and population decline. Japan precedes other countries in experiencing a "super-aging" society not only in rural but also in urban communities. Japan's experience could provide lessons from which other countries might learn. Such social experiments could inform other countries, especially the Asian countries. Japan's new community-building efforts present examples for other countries that seek to strengthen social relationships. High labor force participation among Japanese older adults can provide insights for aging societies (Ichimura et al. 2009). Healthcare networking will play the important role as pathway between residents required healthcare and residents required strengthening community coping capacities against disaster.

6.9 Conclusion

Disaster impact people not only physically but psychologically, socially, and economically. Community-based disaster measures are required to respond during reconstruction after a disaster and to prepare for the next disaster. However, as only conventional disaster measures will make it difficult for local residents to be aware of disaster prevention, this study proposed some indirect disaster prevention measures. In particular, healthcare is important for victims and necessary for people. Hence, disaster measures to strengthen the community through healthcare and utilizing first-aid training will be needed in the future under a fast-aging society, as further aging and a declining birthrate is expected in the near future. It is necessary to strengthen the community network through a daily healthcare network with cooperation among social welfare, medical and local government agencies at multi levels (individual, community, and social) to maintain good healthcare for local residents after disasters, and to promote health.

Acknowledgements The authors acknowledge the financial support of CWS and MERCY Malaysia to conduct fieldwork in Tohoku disaster affected areas.

References

Axelrod C, Killam PP, Gaston MH et al (1994) Primary health care and the midwest flood disaster. Public Health Rep 109(5):601–605

Emergency and Humanitarian Action (2011) From relief to recovery, The WHO Tsunami Operations. http://www.searo.who.int/en/Section1257/Section2263/Section2304_13176.htm. Accessed 15 Dec 2011

Hygo Prefecture (2005) The investigation of death in the Hanshin-Awaji earthquake. http://web.pref.hyogo.jp/pa20/pa20_000000016.html. Accessed 27 Nov 2013

Jackson M (2004) Volunteers boost ambulance crews. BBC News (BBC). http://news.bbc.co.uk/1/hi/health/3604976.stm. Accessed 27 Nov 2013

New York State Health Foundation (2013) Community Health Workers. http://nyshealthfoundation.org/search/results/search&site=site_1&keywords=community+health+worker. Accessed 27 Nov 2013

Presno LC, González CL (2007) Family medicine in Cuba: start, present, and future. Aten Primaria 39(5):265

Report to the Prime Minister of the Reconstruction Design Council in response to the Great East Japan Earthquake. Towards Reconstruction, "Hope beyond the Disaster" (2011). www.cas.go.jp/jp/fukkou/.../report20110625.pdf

Wachtendorf T, Tierney KJ (2001) Disaster resistant communities initiative: local community representatives share their views: Year 3 Focus Group Final Project Report. Report to the Federal Emergency Management Agency. Disaster Research Center, University of Delaware. Available at http://www.udel.edu/DRC/YR3.pdf

Chapter 7
Roles of Community Radio in Disaster Management: Reflections from Japan

Jun'ichi Hibino and Rajib Shaw

Abstract When a major disaster occur in Japan, community radios play the role of disaster radios, or emergency disaster radios are set up in the affected area. In this way, victims can obtain the information they need to overcome the disaster. What's more, besides what has already been said, the role of community radios in disaster management is not confined to providing disaster information only. The presence of the community radio in every phase of a disaster; 'disaster mitigation', 'preparation', 'early warning', 'response', 'recovery and revitalization', is essential for the exchange and sharing of information and dialogue among residents, as well as the enhancement the community's capability and of self-government ability. That is, the community radio is the ideal tool because it has deep roots in the community, has built bonds among the people, and maintains a community identity.

Keywords Community radio • Earthquake • Hanshin-Awaji • Natural disaster • Recovery • Temporary disaster radio • Tohoku • Tsunami

7.1 Introduction

Recent disaster management paradigms consider all efforts to be part of a mitigation cycle (emergency response—recovery). Two of the most important elements within this cycle are information and communication, both of which are extremely influential in reducing disaster risks. Information and communication frameworks that

J. Hibino
FMYY, Kobe, Japan

R. Shaw (✉)
Graduate School of Global Environmental Studies, Kyoto University,
Sakyo-ku, Yoshida-Honmachi, 606-8501 Kyoto, Japan
e-mail: shaw.rajib.5u@kyoto-u.ac.jp

R. Shaw (ed.), *Community Practices for Disaster Risk Reduction in Japan*,
Disaster Risk Reduction, DOI 10.1007/978-4-431-54246-9_7,

raise community awareness must be put in place before disaster strikes, so that residents can be more responsive to their vulnerability and threats of disaster. If a disaster does occur, early warning and emergency response systems created through the benefits of information and communication technology can play an important role in disaster mitigation. After the disaster, information and communication continues to be needed in recovery, reconstruction and rehabilitation (Shaw et al. 2012).

Information and communication have not been fully developed as the main part of disaster management. Most practices consider it as a supporting system that has not been integrated into comprehensive disaster management. In fact, information in the field of disaster management is a basic right of every citizen. Quick, accurate and appropriate information can ensure people's ability to get shelter and survive a disaster. However, judging from a few experiences in disaster situations in Japan, there have been some instances in which information and communication technology have been used to support disaster management. It is interesting that some communities have utilized a form of information and communication technology that is close to their daily lives: community broadcast radio (Fraser and Estrada 2001; Kanayama 2007).

The use of community broadcast radio, more simply called "community radio," shows that the radio can play an important role in carrying out disaster management efforts. Community radio is often more accepted by its surrounding communities because its programs are made in accordance with local social and cultural conditions.

7.2 On-Air Activities of Community Radio

There are high expectations being placed on community radio stations as a support tool for relief and recovery activities, where there are many natural disasters such as earthquakes and tsunami. Beginning with the Hanshin-Awaji Earthquake in January, 1995, the Sumatra Earthquake and Tsunami in December, 2004 and the East Japan Earthquake and Tsunami in March, 2011, the enormous role community radio stations played in emergency relief for the weaker communities gained attention.

The four goals of community radio station in disaster management are: (1) the circulation of emergency information (disaster warnings, evacuation advisories, etc.) to secure the safety of disaster victims, (2) providing relief and support related information of local authorities and NGOs, (3) providing information related to recovery plans in the disaster-hit area and promoting the exchange of residents' opinions, and (4) contribute to maintaining the mental and physical health of disaster victims (Shaw et al. 2012).

In accordance with these goals, the following are the types of programs that are broadcast: (a) providing information, (b) citizen participation, (c) healing mental trauma, (d) monitoring activities. Features of these programs are as stated below.

7.2.1 Providing Information

The type of programs which occupy the largest amount of program time are programs that provide disaster related information stated below. Sometimes staffs of community radio do not directly pass on this information, staff from administration authorities and NGOs which are the source of the information appear on the radio to pass on the information. However, unless the staff of radio community stations and staff of the administration and NGOs build a good relations on a daily basis, and unless community radio stations do not broadcast information from the administration and NGOs on a daily basis, cooperative activities will not function when a disaster occurs (Kanayama 2013).

- Disaster information (disaster warnings and evacuation advisories)
- Safety/contact information
- Aid/support information (volunteer center support information for disaster victims)
- Administration authorities information (explanation of administrative application procedures)
- Recovery of essential utilities (electricity, gas, etc.)
- Living information (information on opening of temporary shops, etc.)

For immigrant minorities who experience language barriers, information should be provided in the language they normally use.

In Japan, to be prepared for a disaster, there are many cases where the local government and the private sector set up low-power output radio stations. There are not a few local government offices that have a broadcasting facility for use when a disaster occurs in the office building. Through an agreement between the local authority and community radio station, the local authority can broadcast directly from the local government office (Hibino 2011).

7.2.2 Citizen Participation

On community radios, by groups and residents in the disaster-hit area as well as support groups from outside appearing on live radio programs or on interview programs, their activities will be known to a wider audience. This also promotes mutual understanding among people in different positions and of different opinions. Also, disaster victims can also have the opportunity to pass on how they feel and what they think and by doing so give themselves encouragement.

7.2.3 Healing Mental Trauma

The radio is a media that communicates to its listeners through sound. Fond, old songs and music, local folk tales and folk songs, comedy, messages from priests, monks, etc.; it is the duty of community radio in the disaster-hit area to broadcast programs that through voice, music, etc. ease disaster victims' suffering, even a little. These kinds of programs are often aired at night or between information programs. If disaster victims listen to disaster related information only, they become depressed. The goal of this type of program is to try to take the listener away from the serious and gloomy reality and soothe their aching hearts.

In the 2011 Great East Japan Earthquake and Tsunami, more than 20 low-power output temporary disaster radio stations broadcast in the local dialects, and broadcast local folk songs and folktales together with disaster information. Such broadcasts soothed the minds of disaster victims in shelters, especially at night when there was no electricity (Murakami 2012).

7.2.4 Monitoring Aid/Support Activities

A major role of the community radio is to confirm if the activities of the administration and NGOs in the disaster area meet the needs of the victims and if the promises made to the victims are being carried out and pass this information on to the residents. Having programs that invite staff from administrative authorities, disaster prevention agencies and NGOs on programs to answer questions from residents can improve the quality of activities in the disaster-hit area. For the planning and implementation of the content of the above programs, it is very important to reflect the situation and opinions of a wide variety of residents in the disaster-hit area, and have many residents participate in programs and management and make the radio station as open as possible to many residents. Furthermore, in the disaster-hit area of the Great East Japan Earthquake and Tsunami, two temporary disaster radio stations broadcast live a municipal assembly and town assembly. In the assemblies, various measures from the relief, restoration to the revitalization periods were discussed and determined. Although the assemblies are open to the public, residents in the disaster stricken area could not go to listen to them. However, they could listen to the content of the assemblies on the community radio. In this way, it is hoped that the policies and activities debated in the assemblies meet the needs of the residents (Rikuzen Takada Saigai 2013; Ringo 2013).

7.2.5 Passing on the Experience of Disaster

It can, on an everyday basis, question how it should act in emergencies. No matter where the disaster-hit area might be, if residents had even a little knowledge about natural disasters and the damages caused by disasters, and the possibility many lives might be saved. It is almost impossible to prevent natural hazards, though, we can reduce the damage they caused by learning how to respond. What community radio can do is to increase resident's knowledge of disasters risk reduction, and pass it on in as many ways as possible to people and society.

Although it is important to acquire knowledge about disaster risk reduction from experts, there are many things that can be learned from residents who have actually experienced a disaster. Every resident can look back and see 'what was I able to do' and 'what was I not able to do' and form these lessons can be gained. Everyone has their own story which should be shared with the community so that the lessons learned from it can be made use of when the next disaster occurs. This activity is referred to as passing on the heritage of disaster experiences through storytelling. Radio FMYY, community radio station in Kobe, has continued this broadcasting activity from immediately after the Great Hanshin-Awaji Earthquake.

In order to prevent the experiences of the Great Hanshin-Awaji Earthquake from being forgotten, a 30-min program "**Daishinsai wo kataritsugu**" (**Continuing to talk about the experiences of the great earthquake disaster**) is broadcast every week on Sundays (holidays) by FMYY. In this program the community passes on to future generations the important themes of "The Preciousness of Life" and "Lessons Learned from the Earthquake," and create communities that can cope with natural disasters by having disaster victims, disaster relief volunteers, local government staff and specialists, etc. discuss the earthquake disaster and how to create disaster-strong communities. The program is uploaded onto the Internet with the script following day and can be downloaded on demand by those who missed it. The scripts are translated into English, Korean to expand the knowledge and the wisdom across the country. And, with the assistance of FMYY, community radio stations around Mount Merapi volcano in Central Java, Indonesia have begun the same activity of broadcasting disaster victims' stories in order to pass them on (FMYY 2013a).

7.3 Off-Air Activities of Community Radio

As the community radio shifts from giving information on emergency water supplies and open-air kitchens early after the disaster to providing recovery information and information to cheer up and encourage residents, it becomes a radio station that supports the recovery and revitalization of the disaster-hit area. However, by just broadcasting programs from the studio, it cannot be said to be a radio station close to the people. There are a variety of roles the community radio can play beyond broadcasting.

As Mr. Zane Ibrahim (founder of Bush Radio), known as the father of South African community radio, said, "community radio is made up of 90 % community, 10 % radio"; community radio uses the radio for community creation. In the same way, community radio uses the radio for support activities and recovery/revitalization activities.

In many disasters in the past, people who formally had no connections were, through participation in community radio station activities, able to make connections with people like themselves. Thus through disaster victims' mutual support activities, holding discussion meetings for recovery and recovery events, the basis for self-determining citizen activities in the community is created. The community radio has the function of encouraging disaster victims to participate in community activities (Shaw et al. 2012).

In Koriyama City, Fukushima Prefecture, many of the residents of Tomioka Town have been forced to evacuate because of the accident at the Fukushima Daiichi Nuclear Power Plant in March 2011. In the temporary housing complex in Koriyama City there is an assembly center where residents get together which has a community radio. Care activities are conducted in the center to relieve the physical and mental exhaustion caused by living in prolonged evacuation. Staff of the community radio station hold karaoke competitions at the center, work in the fields with the disaster victims and help the victims find a aim in their life while in the temporary housing.

7.3.1 Synergetic Effect with Internet

In the first place, the community radio is a tool to work to improve the quality and spread the activities of people who are working to solve social problems in the community. This is also the reason why groups doing support activities in disaster-hit areas set up community radio stations.

Also, the community radio station plays the role of connecting people within and outside the broadcasting area. With the Internet, it is possible for people outside the disaster-hit area to listen to community radio programs where they can get to know detailed information that the mass media does not give, as well as hear the voice of the disaster victims. Also, this information is more reliable and trustworthy than that found on social media networks like ‘twitter’. There are not a few cases where the community radio motivated people from outside the disaster-hit area who listened to their broadcasts to participate in support activities in the disaster-hit area.

After the Great East Japan Earthquake and Tsunami, special disaster FM stations not only broadcast on low-output radio wavelengths, they also simultaneously broadcast on the Internet reaching areas outside the disaster-hit area. As time passed from the disaster, the mass media gradually reduced broadcasts and people’s concerns about the affected area gradually faded. However, community radio stations in the area continued to broadcast the current daily situation in the area. And through Internet broadcasts, it was possible to fill in the awareness gap between inside the affected area and outside (Murakami 2012).

In order to convey information about and from Tomioka Town to Tomioka residents who have sought refuge throughout the country, Tomioka Town community radio stations distributed mobile tablet devices connected to the Internet free of charge. With these tablet devices, all programs can be heard by on demand broadcasting, and listeners who have heard these programs can participate in exchange forums using the devices.

On the other hand, as time passes, the difference in interest and concern between within and outside the disaster-hit area grows. Even though the disaster-hit area still needs support from the outside, as mass media coverage decreases little by little, people’s interest gradually weakens. In the course of time, the community radio that becomes the recovery radio and eventually the community radio is made possible through the people who met because of the disaster.

7.3.2 Holding Memorial Events of Disasters

As an off-air activity, events play an important role in providing listeners with an opportunity to learn with their five senses about disaster prevention and about the importance of life. Unlike watching TV or listening to the radio, participation in events creates opportunities to actually experience many things. Also, informing TV and major radio stations in advance that the community radio will broadcast an event live on the theme of natural disasters, and broadcasting the event live will result in many listeners gathering and participating. Broadcasting events in which residents participate, gives listeners a sense of participation and a feeling that it is some-thing close them.

For example, through live broadcasting of annual memorial events of natural disasters that are held together with local residents, we can pass on to listeners experiences of disasters, the message of the preciousness of life, and the importance of disaster prevention activities. Another event representative of a disaster prevention event is the 'town-walk workshop.' It is an open event where community radio's listeners can listen to disaster prevention specialists, program personalities and announcers as they walk around the town looking at it from a disaster prevention perspective. It is an event combining disaster prevention study and hiking. Live coverage of the event with commentators and announcers gives listeners who cannot participate the chance to learn about things that are useful in times of disaster or problems that might occur in a disaster. In this way, events provide radio stations with the opportunity to bring listeners directly in touch with disaster damage risk reduction. Also, live broadcasting of these events can give listeners the feeling they are actually at the event (SEEDS Asia 2009).

On the anniversary of the Great Hanshin-Awaji Earthquake (January 17), the commemorative event 1.17 Kobe ni Akari wo in Nagata (1.17 "Bring Light to Kobe" in Nagata) is held together with many listeners and local residents as a memorial to the victims of the earthquake as well as to increase disaster prevention awareness. The figures "1.17" are created with candles which are lit at the time the earthquake struck. At that time all participants offer a silent prayer for the repose of the spirits of earthquake victims. The event includes the singing of a song born from the earthquake by junior high school students, a Japanese drum performance in memory of the victims and for the recovery of the disaster-hit area and music performances on stage. A community radio station called "Radio FMYY" broadcasts live radio coverage of these memorial events as well as talks about earthquake experiences and disaster prevention activities with earthquake victims, specialists, local government staff, NGO staff and various artists from its outdoor studio set up at the event venue. However, as time passes, the number of these events has decreased. In order to prevent the experiences of the earthquake disaster from being forgotten, we produce programs that pass on the lessons learned from the earthquake to enhance residents' disaster prevention awareness, and provide other community radio stations with these programs. As well as this, we provide live coverage of events in the disaster-hit area on January 17 by telephone to a number of radio stations (FMYY 2013b).

7.3.3 Disaster Management Audio Materials for Community Radio Broadcasting (DMAM)

In the Asia-Pacific region where there are many disasters such as earthquakes and tsunami, great expectations have been placed on community radio as a powerful tool to foster rescue and rehabilitation activities. The most advanced illustration of this was in the Great Hanshin-Awaji Earthquake in 1995, and the Sumatra Earthquake and Indian Ocean Tsunami in 2004. In November 2006 at the workshop "Community Radio and Disaster and Recovery" at the ninth World Conference of World Association Community Radio Broadcasters (AMARC), the important role of community radio stations played in these tremendous disasters was reported. This provided the first opportunity to share experiences for community radio stations of the world. The next step was to fulfill the need for development of training and contents based on their experiences. The answer to that need is Disaster Management Audio Materials for Community Radio Broadcasting (DMAM).

The DMAM was created by the Disaster Reduction Learning Center (DRLC) which was jointly established by JICA (Japan International Cooperation Agency) and Hyogo Prefectural Government. It was developed to be used as an educational material for JICA Learning Programs. And at the same time, this tool will help community radio stations and local communities to strengthen their disaster preparedness, providing them with easy access to disaster management audio materials. This tool contains, on a single CD-ROM, 193 audio and text messages in nine languages (English, Chinese, Thai, Tagalog, Indonesian, Vietnamese, Portuguese, Spanish, Russian) for broadcasting on community radio stations when four types of disasters (earthquakes, tsunami, landslides and floods) occur. It offers the user easy operation: simply insert the CD-ROM in a computer and start the Web browser. Community radio stations can easily provide disaster related information to their listeners and local communities (Sumratne and Hibino 2010).

DMAM can be also used for disaster reduction activities in everyday life. DMAM is not only for community radio activity use. The audio materials can be used by regional resources, such as people involved in disaster risk reduction, the community, NGOs, schools, etc., as well as in workshops and disaster reduction education. The audio materials are just that, materials. Because they are materials they can be used freely and creatively. They can also be altered in many ways to suit the culture and customs of each country and region.

7.3.4 Community Workshop on Utilizing DMAM

The role of community radio does not stop at emergency response and rehabilitation. In normal times, it is important that the role of community radio be a channel for disaster mitigation education for the communities. It is very important that this process is done in disaster-prone areas. Therefore, when a disaster occurs, people will be better prepared to deal with it. Realizing the importance of disaster mitigation education for the community, JICA Hyogo/Disaster Reduction Learning Center (DRLC), Combine Resource Institution which is a local NGO in Jogjakarta, Indonesia to support community radio stations, and Radio FMYY held a workshop entitled "Community Based Disaster Management Utilizing Community Radio" on 2–5 August 2009, in two community radio stations around Yogyakarta, which were named Lintas Merapi FM on volcano disaster and Angkringan FM on earthquakes and floods.

Prior to this, JICA Hyogo/DRLC and FMYY had produced audio materials containing practical guidelines in dealing with natural disasters, like earthquakes, tsunami, landslides and floods. These two community radio stations were asked to listen to the audio products and then later to adapt it to the needs of the local communities. This workshop was very important for both stations because it was the first time they could hold a special gathering with residents to discuss the roles of community radio in disaster education for the community. Although the audio materials does not include volcano disasters, it has inspired them to produce the same kind audio materials. They were very eager to make an audio product concerning the signs of volcanic eruption. In order to get correct information, volcanologists was consulted for the product. Eventually such information was packaged using the local language and a style that the people could understand.

In this workshop the community radio personnel and the people also developed some new ideas. For example, the traditional singing group that has a special program on Angkringan FM was encouraged to compose songs talking about early warnings for dealing with any earthquake. In the workshop, local idioms became the primary choice in stirring up public awareness of disaster mitigation. This workshop also encouraged the community radio stations to expand its roles not only in radio broadcasting but also in off-air activities like children's painting contests with the theme of disaster, workshops in schools, etc. (JICA 2010).

Considering vulnerable areas in many countries to various forms of natural disasters, such as earthquakes, volcanic eruptions, landslides, hurricanes, floods, tsunamis, etc., the model that has been developed in this workshop should be applied in all community radio stations. Hence, community radio can be the spearhead of disaster risk reduction education through both on-air and off-air programs. The closeness of the community radio station to its supporting communities can provide the power to play a role in reducing the risks of disasters in the future.

DMAM can act as a useful tool for community radio stations in their role of promoting disaster management in communities, and with some creativity and planning it can be used as a new basis to promote disaster management activities to meet various situations and experiences of each region or community (FMYY 2013a).

7.3.5 Networking with Other Radio Stations

However, there are limits to what one radio station can do during any of the following phases: pre-disaster prevention, during disaster, and the post-disaster reconstruction period. For example: What would happen if the building housing the radio station was seriously damaged in a natural disaster? What would happen if radio station staff were victims of the disaster and couldn't carry out the broadcasting work? In situation like this, the most dependable measure is support from other stations. Because they know about radio stations and how they are run, they can provide various kinds of support necessary for broadcasting. However, if there is no regular contact or communication among stations, mutual support will not work efficiently when a disaster occurs. If stations meet regularly, co-produce radio programs, and work together, a disaster management network can be built. Through regular meetings, not only for disaster risk reduction but also for exchanges about everyday affairs, a mutual exchange of knowhow and experience can be created (SEEDS Asia 2009).

There are three community radio stations on Amami Oshima island, where torrential rain is common. When the 2010 heavy rain disaster happened, the radio stations announced who was sheltering in which evacuation shelter. This helped to prevent secondary disaster. Also, neighboring radio stations cooperated and broadcast each other's information from their station. However, this cooperation was a coincidental occurrence resulting from the disaster (Nagasaka et al. 2011). Learning from this, three community radio stations and the local authorities on the island began to hold regular meetings for coping with disasters. In these meetings, a structure was set in motion to pass on trustworthy information as quickly as possible to residents in times of disaster through the cooperation of community radio stations and local authorities and among community radio stations themselves.

7.4 Conclusion

While it is true that the mass media contributes to society by disseminating information, it is also true that it is not a medium that can meet the individual needs of a diverse society. This is not only true on a daily basis, but also true in times of a disaster. In Japan, the major role community radios play in disasters became visible when the Great Hanshin-Awaji Earthquake struck in 1995. At that time, temporary disaster radio stations were set up over a wide area in disaster hit Hyogo Prefecture. This was the first time temporary disaster radio stations were set up to pass on disaster information. However, because the broadcasting area was so wide, detailed information which was not easily obtained from the mass media which victims needed was not adequately provided.

On the other hand, at the same time, a volunteer group set up a disaster radio station without obtaining a license. This radio station was a low-power output community radio targeting the local community and by providing detailed disaster information gained the great trust of the victims. One year later, the station obtained a community radio license. Thereafter, when a major disaster occurs, community radios play the role of disaster radios, or emergency disaster radios are set up to target communities. In this way, victims can obtain the information they need to overcome the disaster.

What's more, besides what has already been said, the role of community radios in disaster management is not confined to providing disaster information only. The presence of the community radio in every phase of a disaster; 'disaster mitigation', 'preparation', 'early warning', 'response', 'recovery and revitalization', is essential for the exchange and sharing of information and dialogue among residents, as well as the enhancement the community's capability and of self-government ability. That is, the community radio is the ideal tool because it has deep roots in the community, has built bonds among the people, and maintains a community identity. This is especially so in the phase from the disaster to recovery, it should not only be the opinion of the loudest voice, it is the role of the community radio to shine light on those on the outskirts of the community whose existence is not seen and should not be forgotten. That is not all that the media can do. It can, on an everyday basis, question how it should act in emergencies. No matter where the disaster-hit area might be, if people had even a little knowledge about natural disasters and the damages caused by disasters, and the possibility many lives might be saved. It is almost impossible to prevent natural hazard, though, we can reduce the damage they caused by learning how to respond. What community radio can do is to increase resident's knowledge of disasters, and pass it on in as many ways as possible to people and society.

Sadness, suffering and bitterness often make a community shorten its memory of an event. The people want to forget their beloved families and relatives who died in a disaster. Such a struggle is certainly very understandable in their effort to survive the past and go on to the future. But the process of forgetting is also critical, because it can cause them to become unaware of a disaster that someday may strike again. This is where a community radio station can contribute in most a delicate and elegant way to notify them that past disaster shall not be forgotten, but will be accepted, learned and remembered, so that the same misfortune will not happen again in the future.

References

FMYY (2013a) http://www.tcc117.org/fmyy/index.php?e=1928. Consulted on 15th July 2013

FMYY (2013b) http://www.tcc117.org/fmyy/index.php?e=315. Consulted on 30th July 2013

Fraser C, Estrada S (2001) Community radio handbook. UNESCO, Paris, Available at http://developingradio.org/files/UNESCO%20CR%20Handbook.pdf

Hibino J (2011) Community radio to promote social inclusion of minority- management system to sustain activities of radio. FMYY, Kouyo Syobou

JICA (Japan International Cooperation Agency Hyogo International Center) (2010) Handbook for the utilization of DMAM (disaster management audio materials) for community based disaster management. JICA Hyogo/Disaster Reduction Learning Center. Available at http://www.drlc.jp/english/?p=409

Kanayama T (2007) Community media—community FM ga chiiki wo tsunagu. Keio University Press, Tokyo

Kanayama T (2013) Community radio and the Tōhoku earthquake. Int J Jpn Sociol 21(1):30–36

Murakami K (2012) An overview of the disaster information transmission system of the local governments in the post-quake Japan. Research for broadcast, NHK Broadcasting Culture Institute. Available at http://www.nhk.or.jp/bunken/summary/research/report/2012_03/20120303.pdf

Nagasaka T, Tsubokawa H, Lee T, Suzuki H, Kioshita K, Amano T (2011) The disaster response in Amami Heavy Rain on October 2010. Disaster Prevention System Research Center, National Research Institute for Earth Science and Disaster Prevention, Japan. Available at http://dil-opac.bosai.go.jp/publication/nied_natural_disaster/pdf/46/46-02.pdf

Rikuzen Takada Saigai FM (2013) http://rikuzentakata-fm.blogspot.com/2013/06/613.html. Consulted on 25th July 2013

Ringo Radio (2013) http://ringo-radio.cocolog-nifty.com/blog/2013/04/index.html. Consulted on 25th July 2013

SEEDS Asia (2009) Know disaster, tell disaster risk reduction: training handbook for media professionals. Available at http://www.unisdr.org/we/inform/publications/11705

Shaw R, Hibino J, Matsuura S (2012) The role of community radio in disaster. Church World. Available at http://www.preventionweb.net/files/29931_29931radioalllowres1.pdf

Sumratne I, Hibino J (2010) Application of DRR tools: sharing Asian experiences. ADRC (Asian Disaster Reduction Center). Available at http://www.adrc.asia/publications/Tool_Book/PDF/COMMUNITY.pdf

Part III
Case Examples

Chapter 8
Community Development and Disaster Preparedness in a Depopulated Society: Case of Tosashimizu City, Kochi Prefecture

Junko Mimaki and Rajib Shaw

Abstract Japan is known for its vulnerability to disasters. Japan is highly ranked fourth (ranked 16th) in world risk index (United Nations University et al., World Risk Report, 2012). Furthermore, since Japan is located in the boundary between the oceanic plate and the continental plate, it is more geographically vulnerable to damage by earthquakes and volcanoes than other countries [Cabinet Office, Bousaihakusho (White paper on disaster preparedness), 2007a]. And since Japan is surrounded by the sea and has complicated coast lines, the coastal areas, particularly where populations, assets, and city function have accumulated, are vulnerable to tsunamis (Cabinet Office, Bousaihakusho (White paper on disaster preparedness), 2008a). The occurrence of the Great East Japan earthquake (magnitude 9.0) in 2011 is still fresh in our memory. It triggered powerful tsunami waves that reached heights of up to 40.5 m. A considerable number of casualties were reported: 15,878 deaths, 6,126 injured, and 2,713 missing across 20 prefectures. A wide-ranging damage was also experienced: the collapse of houses, destruction of buildings by fire, aquatic submersion of houses by tsunamis, damage to fishery harbors by castaways, and more. Though the Japanese government has promoted community based disaster preparedness in collaboration with local governments, communities in rural areas have gradually faced society's rapid aging and depopulation. For some communities, it has become more difficult to implement even a communal event. In such a case, what should be paid attention to promote community based disaster preparedness by a local government? This chapter focuses on a Japanese rural community categorized as "a marginal village", of which communal activities stagnated at the commencement of community based disaster preparedness. This

J. Mimaki (✉)
Japan International Cooperation Agency (JICA), Tokyo, Japan
e-mail: Mimaki.Junko@jica.go.jp; mimaki.junko@gmail.com

R. Shaw
Graduate School of Global Environmental Studies, Kyoto University, Kyoto, Japan
e-mail: shaw.rajib.5u@kyoto-u.ac.jp

R. Shaw (ed.), *Community Practices for Disaster Risk Reduction in Japan*, Disaster Risk Reduction, DOI 10.1007/978-4-431-54246-9_8,

study was conducted through field surveys in the length of 8 years. It attempts to clarify key factors to promote communal activities for disaster preparedness and to show that it is indispensable for local governments to promote them in the perspective of community development in order to induce the capacity of a community for disaster preparedness and enroot that in the community.

Keywords A marginal village • Aging society • Community development • Social capital • Stagnation

8.1 Introduction

In Japan, after the Kobe earthquake (in 1995), there has been an increasing awareness of the importance of the social aspects of disaster preparedness, particularly the community's capacity to respond. The Japanese government amended the "Basic Policy" in the "Basic Act for Disaster Countermeasures" and added the importance of communication and participation between residents and local government as well as among the residents themselves. The Japanese government, in collaboration with local governments, has promoted establishing a Voluntary Disaster Preparedness Organization (VDPO) to improve community-based disaster preparedness. The rate of organization is 75.8 %, as of April 2011 (Nara Prefecture 2012). Perry and Lindbell (2003) pointed out that smaller communities are largely reliant upon informal, personal relationships for risk identification, assessment, and reduction. Previous research (Mimaki et al. 2009) revealed that VDPO activities contributed to raise residents' awareness of disaster preparedness as well as deepen bonds in the community. The survey showed that VDPO contributes to improving residents' awareness of disaster preparedness (perception of risk, self-help, and mutual help). Moreover, it implies the existence of some problems to be paid attention to. Although activities for disaster preparedness through VDPO provided residents with proper recognition about evacuation and aroused residents' interest in disaster information, the activities did not urge the residents to take action toward countermeasures such as stabilization of furniture. The result also indicates that collective action through VDPO contributes to deepening communal bonds. That is, participation in VDPO might be an opportunity for residents to reconsider their identification as a member of the community. It is therefore necessary to take measures in the activity of VDPO to encourage residents' to act.

Recently, beyond natural vulnerability, it is said that social factors such as depopulation and aging have increased in rural areas, posing further risks. The period of high economic growth (1960–1973) resulted in population flow from rural areas to cities (Shinohara 1991). "Depopulation" refers to residents' difficultly in sustaining their quality of life, due to a decrease in population. If the number of residents decreases, it is difficult for the community and local government to continue fundamental services such as disaster preparedness, education, and access to health care (Imai 1968). Yamazaki (1994) indicated that "depopulated communities are easily exposed to damages caused by disasters". In addition, disasters accelerate depopulation and disruption of communities. When the population of communities in rural areas decreases, their residents rapidly age. In the 1980s, Japan's aging population rate (people older than 65) was mid-level among developed countries. Japan is now at the top among them. In 2007, the aging population rate was 21.5 % and is expected to reach 40.5 % by 2055. Though the aging rate has been low in three major metropolitan areas (Tokyo, Osaka, and Nagoya), the rate in other areas has been high. Shimane Prefecture (28.2 %) tops the list of the aging population rate, followed by Akita Prefecture (28.0 %), and Kochi prefecture (27.2 %) (Cabinet Office 2008b). It follows that the urgency of disaster preparedness in rural coastal communities has increased. At the same time, it implies that it has become more difficult to continue communal activities, including VDPO activities, in these areas.

Under these circumstances, how then should VDPO activity be promoted to build a resilient community, considering the effect of VDPO mentioned above, in case a community faces social stagnation and does not have sufficient capacity to implement VDPO activities such as evacuation drills? This chapter examines key factors which local government should pay attention to in order to promote VDPO activities in rural and aging society. The chapter is presented in three sections. The first section (Sect. 8.2) contains an overview of the study area. The second section (Sect. 8.3) describes a case study on a community in Tosashimizu city. The final section (Sect. 8.4) presents some conclusions and recommendation for further research as this article is an authors' personal opinion it does not correspond with statement of authors organizations.

8.2 Overview of the Study Area

8.2.1 Rationale for the Selection of the Site

This study is focused on Tosashimizu city, Kochi prefecture, Shikoku region, Japan. The rationale for the selection of the site is as follows.

8.2.1.1 Necessity to Enhance the Community's Capacity for Disaster Preparedness

Tosashimizu city has been hit frequently by natural disasters. The effects of these disasters can become worse if communities are not prepared. In 2001, due to lack of forest management in the area, some communities were affected by flooding caused by pieces of drifted wood. Recently, the city is predicted to be seriously damaged by a Nankai earthquake within 30 years. It is estimated that over 70 % of the whole area of the city will be inundated by tsunamis. In 2012, the Cabinet Office announced officially that the expected tsunami height will be 34 m in the worst-case scenario. Accordingly, the Tosashimizu city government has officially been promoting the establishment of VDPO since 2003.

8.2.1.2 Depopulated Area and High Aging Rate

The population of the city has decreased due to the decline of primary industry and influx of population to large cities during the period of high economic growth (1960–1973). The ratio of aging in Tosashimizu city (34.6 %) (Tosashimizu city 2005) is higher than the national average (20.1 %) (Cabinet Office 2007b). In turn, the city is facing various difficulties caused by its aging community.

Tosashimizu city is located about 170 km southwest of Kochi city (prefectural seat) and is the southernmost city of the Shikoku region. It is one of the cities in Kochi prefecture, Japan. The population of Tosashimizu city is 16,621 (census 2010) (Tosashimizu city 2013). It is surrounded by the sea and was known for deep-sea fishing (bonito). At present, it is known for its coastal fishery and sightseeing. Owing to its geographical location, the city is frequently hit by typhoons.

8.2.2 Local Governments' Policy and Measures for Community Based Disaster Preparedness

In order to reduce the vulnerability to damage caused by earthquakes, the Kochi prefectural government has been promoting the establishment of VDPO and has aimed to establish it in each of all the communities in Kochi prefecture. The prefectural government decided, in collaboration with local governments, to provide financial support for VDPOs which were established in the fiscal year of 2008. Hence, the Tosashimizu city government has encouraged communities to establish VDPO since 2000. At normal times, VDPO is expected to play the role of educating residents about disaster preparedness such as implementation of evacuation drills and so on. As a rule, set by the city government, all VDPOs are obligated to implement their first evacuation drill within the same fiscal year. After the establishment of VDPO, the city government leaves the VDPO to create its activities and manage itself. The city government considers VDPO a communal organization, bearing its own responsibility.

8.3 Case Study

8.3.1 General Feature of the Kainokawa-ura Community

In this part, the example of the Kainokawa-ura community in Tosashimizu city is considered. The rationale for the selection of the site is as follows:

- Depopulated area and high aging rate
- Location (community located in the coastal area and has tsunami risk)
- Stagnation of community activity (at the establishment of VDPO) and Continuation of VDPO activities (as of January 2013)

The Kainokawa-ura community is located in a coastal area, about 20 km east of the nearest downtown area. It is famous for its fisheries, particularly fishing using fixed nets. Recently the number of fishermen has decreased due to aging and economic depression. According to the leading community representative (hereafter, "leader"), in the Meiji era (since the latter 1860 until the early 1910), the population of the community was 300–400, and the economy was centered on fisheries. Since the 1960s, however, the population has started decreasing. At present, the population is 121 (April 2007 census). Recently, the population has become an aging society, with the aging rate (ratio of population over 65) of 53.45 %. The community is categorized as a marginal village. Owing to depopulation, Kainokawa Junior High School (established in the Meiji era) and Nursery School were closed in 2007, and in 2009 Kainokawa Elementary School was closed. As it is located in the coastal area, it has a risk of tsunami. In 2001, the community was partially damaged by flood.

The authors implemented key informant interviews with the local government officials, the community leader, the board members and residents from 2004 until 2012.

8.3.2 *Communal Management and VDPO Activity*

Although a VDPO was established in 2004, the activity of the VDPO as well as other communal activities stagnated at the beginning. The following are the recordings of communal management and the VDPO categorized into three stages: a stagnant stage (until 2006), a turning point (2007–2008), and an active stage (2009–2012).

8.3.2.1 Stagnant Stage (Until 2006)

The leader has been in charge of community responsibilities since 1992. He worked in local government for about 40 years, and with such a background he was elected as leader with the expectation that he could play the role of a "bridge" between the community and local government. For a long time, the leader believed the community lacked sufficient human resources, and thus he needed to find people who could assist in community events. Furthermore, when the leader encountered a problem, he typically attempted to solve it by himself. In 2006, there was only one communal organization still operating in the community, a senior citizens' group. Though residents over 65 years old are qualified to enter the group, most members of the group are over 80 years old. As for previous organizations, "Seinendan" (Young Generations' Group) was dissolved about 30 years ago, and a women's group was disbanded about 20 years ago. When implementing communal events such as community festivals, the leader had to find citizens to assist and participate in the community. Even with the leader's effort, the number of participants in traditional community events decreased during this period, which consequently brought difficulty in implementing community events.

There were other problems the community began facing. One of which was its relationship with other communities. The Kainokawa-ura community borders the Kainokawa-gou community, which is an agricultural village. Therefore, it is said that the character of each of these two adjacent communities is different (e.g., Kainokawa-gou is more conservative and reserved). Both communities traditionally had a "competitive spirit". But the opportunity to do something jointly was rare. What the residents did jointly was only organizing drinking parties after school athletic meetings and graduation ceremonies. This was because the children of both communities had to attend the same public schools (elementary and junior high). However, in 2007, Kainokawa Junior High School was closed, and currently all students have to attend Shimokawaguchi Junior High School, which is located in another district. Since the closure of Kainokawa Junior High School, there was no longer any drinking party. In an interview done in 2006, the leader said that after the closure of Kainokawa Elementary School in 2009, the two communities might not have further opportunities to have joint parties. Except for the celebratory parties mentioned above, there were virtually no other opportunities for communication between the two communities. For example, although they worship at the same shrine, they hold separate spiritual events.

The area is prone to typhoons and other natural disasters. In 1945, it was hit by a tsunami. Since most of the houses were located in relatively higher areas, the overall damage was not severe. In 2001, the western area of Tosashimizu city, which includes the Kainokawa-ura community, was damaged by flooding; the total amount of the damage was estimated to be over 0.3 billion yen. In 2003, the local government, which was promoting the establishment of VDPO in all communities, asked the community leader to establish a VDPO in the community and gave the leader the guidance manual and an application form to be submitted. In 2004, the application form was accepted and the VDPO was established. The administrative staff of the community concurrently serves as board members of its VDPO. Though the local government has a rule that each VDPO is obligated to implement evacuation drills at least once every fiscal year, the leader hesitated to implement the drills, for the following reasons:

- Reason 1. The leader thought the residents would not agree to perform the drills. Although the Kainokawa-ura community faces the sea, its elevation is relatively high, and thus the leader thought that the Kainokawa-ura community was safer than the neighboring Kainokawa-gou community. In addition, there was no urgent need to take countermeasures.
- Reason 2. It was a 'burden' for the leader to hold a new community event. It was already difficult to find participants for traditional community events such as a communal festival, and hence the leader thought it would be more challenging to find those who would cooperate in implementing the evacuation drill.
- Reason 3. Administrative division within the community was complicated, with many residents scattered throughout the neighboring community areas. Thus, the leader thought it would be difficult for the Kainokawa-ura community to implement the evacuation drills alone.

In 2005, the neighboring Kainokawa-gou community established their own VDPO. The local government made suggestions to both community leaders several times to implement joint evacuation drills within the fiscal year of 2005. Both leaders then made plans to implement the drills in March 2006, with the leader of the Kainokawa-ura community proposing the local government's ideas at the community's annual meeting. Though the residents accepted the plan at the meeting, the drills were eventually canceled due to schedule conflicts related to funeral services in the community. Subsequently, the community did not have the opportunity to reconsider a plan to implement the drills.

8.3.2.2 A Turning Point (2007–2008)

In 2007, a "change" happened in the stagnated community management. The staff of Tosashimizu City Council of Social Welfare (hereinafter referred as "Council") visited Kainokawa community and proposed that the residents establish a new community circle called as "Fureai-ikiiki Community Salon (hereinafter referred to as 'Salon')". In 2006, the Japanese government reexamined the elderly care insurance system and made a bold policy shift toward reinforcing the emphasis on care prevention. In line with such policy changeover, a nationwide body, Japan National Council of Social Welfare, promoted the establishment of a Salon in each community. The Salon has its worthiness in improving the users 'quality of life and offering safeguards not only to its users but also non-users who tends to isolate themselves from community activities in cases of emergency' (Nakamura 2009). One of the main activities of Salon is calisthenics. This program is introduced to improve the physical health of elderly residents. Following the proposal made by the Council, 15 female residents announced their participation. Their age bracket was from 60 to 70 years old. As they had considered that it was still too early to be a member of the senior citizens' group, they accepted the proposal. Firstly, they participated only in Salon activity (weekly calisthenics program and blood-pressure measurement), but they gradually had come to make a spontaneous offer to support the leader. Then, the leader in collaboration with Salon members decided to warm up their communal summer festival in 2007. They formed the Bon festival dancers. In addition, they offered memorial services to honor members who died within that year. Even if the member died outside the city, they decided, as part of the festival, to provide the service as long as the dead was originally from the community. Many people who were associated with the community came home for the services, and the number of participants of the festival increased. The festival being full with people reminded residents of the community's tradition and halcyon years. Some residents expressed their appreciation for the leader and Salon members' efforts. In the following year (January 2008), three members of Salon were elected as board members of the community.

In respect of VDPO activities, in 2007, the local government proposed again that both Kainokawa-ura community and neighboring Kainokawa-gou community should implement the drills. However, it was postponed due to funeral services. A year later (2008), the local government proposed both communities to implement the drills jointly on 'Municipal Joint Evacuation Drill Day' (September). Both leaders proposed the new idea at the annual meeting of their community, and the residents gave their approval. Both leaders then began holding meetings to discuss how the joint drills would be carried out.

Through the meetings, the leaders shared their problems concerning their communities, encouraged each other, and decided to use the joint drills as an opportunity to "revitalize" their communities. Based on the minutes of the meetings, the leader designated four places, Ipponmatsu, Yasaka Shrine, Minatoyama Field and Ebisu Shrine, as "evacuation sites". In September 2008, both communities implemented the evacuation drills. In addition to the notification done through local government broadcast, the leader informed the residents about the event twice a day (morning and night) from 3 days before the start of "Municipal Joint Evacuation Drill Day" through radio broadcast. The number of participants in the drills from the Kainokawa-ura community was 49, which was larger than the number of participants in other community events. Not only senior citizens but also younger residents who had not actively participated in community events participated in the drills.

Through the implementation of the first evacuation drills (in 2008), some changes were observed in the target community as follows.

Change Concerning Residents: Increase in Residents' Requests for Disaster Preparedness

Before the first evacuation drill in 2008, none of the residents contacted the leader about the necessary countermeasures for disaster preparedness; even the risk of tsunami and earthquake had been broadcasted and reported. After the first evacuation, the residents requested for the promotion of disaster preparedness. Most of the requests were related to the roads designated for use during the evacuation and the evacuation sites. Before the evacuation drills, some residents as well as the leader noticed that some of the evacuation sites, which included shrines, needed handrails to improve their suitability as evacuation sites. After participating in the drills, however, they realized that the stairs leading to the shrines are steep and that they also needed to be adjusted for a gentler ascent. Some residents invited local government officials to the shrine and asked them for official support in remaking the stairs. The local government accepted this request. In October 2008, the local government committed to prepare a bid for redesigning the stairs leading to the shrines. Through participation in the drills, the residents were able to identify necessary countermeasures.

Change Concerning the Leader: Recognition of Communal Possibility and Necessity of Collaboration with Others

The leader was surprised and delighted to see that the number of participants was larger than what he had imagined. The participation of younger generations made the leader realize that the residents needed such activities. At the same time, he "discovered" the existence of communal resources such as the leaders of sub-communities and partnerships with other communities that could be used to resolve community-wide problems. He therefore planned to take appropriate countermeasures that included such communal resources (described below):

- Necessity of preparing emergency rucksacks: In the evacuation drills, young participants carried emergency backpacks holding valuables and goods necessary for evacuation. More elderly residents, on the other hand, did not make such preparations. There are no stores in the community that stock such goods, and thus it is very difficult for senior citizens to prepare such rucksacks. Nonetheless, to help clarify the need for and contents of such backpacks, the leader developed a plan to distribute a list to all residents with illustrations of the goods necessary when preparing for emergency. By using illustrations, in addition to text explanations, the leader was sure that senior community members would understand the list. At the same time, the leader also planned to ask the local government to introduce residents to the ways of procuring the goods.
- Necessity of improving communication with senior citizens: To publicize disaster preparedness information through the community broadcast system, the local government distributed drafts of scripts to each VDPO. However, the leader found the descriptions having "technical terms", which would be difficult for the more elderly residents to understand. To solve this, the leader edited the scripts, paraphrasing them in a language that is easy to understand. Moreover, the leader emphasized the importance of "repeating" keywords, because it is sometimes difficult for senior citizens to understand audio information delivered quickly and heard only once.
- Necessity of involving sub-community leaders: To prompt residents to participate in the drills, the leader devised a plan to distribute a name list of the residents to the sub-committee leaders and asked them to personally call and invite their community members.
- Necessity of preparing for an alternate program for those who cannot participate in the drills: The leader found that there were some residents who could not participate in the drills due to their physical condition, even though they were interested in participating. Therefore, he decided to prepare a separate program for them in the subsequent evacuation drill. Specifically, those residents who were not in good physical condition could participate by simply "standing by at home".

Through the implementation of the evacuation drill, he came to understand how such community-wide drills can be conducted. He in turn received praise and positive feedback from his "collaborator", the leader of Kainokawa-gou community. Such feedback gave the leader a "feeling of confidence" and increased his "motivation for working toward disaster preparedness". One month after the first evacuation drill (October 2008), he experienced lobbying for the first time in his life. In addition, the leader noticed that both communities have "mutual interests and problems" that can be solved through mutual collaboration. For example, in the present study, the leader petitioned local government representatives for support in order to help the neighboring community. Similarly, when the leader of the neighboring community discussed necessary support from the local government, he also asked the local government to support the Kainokawa-ura community.

8.3.2.3 Active Stage (2009–2012)

The leader has held the position since 1992. In 2009, a tertiary communal organization was born. It is called a volunteer group, which is mainly in charge of preparing lunch bags for elderly residents (over 80 years old) once a month. The group is composed of ten residents who are in their 70s. In the same year, a member of Salon who concurrently held the Commissioned Welfare Volunteer proposed the issuance of a communal newsletter to introduce the activities of the three communal organizations. And so a newsletter for the community began with a monthly publication. Though the characteristics of each organization are different, they were energized by their competitive consciousness. After this event, Salon's presence in the community has increased. While they have continued their main activities (weekly calisthenics program and blood-pressure measurement), the group has widened their activities. In fact since 2007, Salon members have been in charge of the dance in summer festival. In 2009, they revived the use of the drum in the dance performance. Many residents commented that they could enjoy the nostalgia of the past when they heard the sounds of the drum. In 2011, the leader, in collaboration with the Salon, held a residents' talent show. In the said event, many residents gave various performances. Furthermore, they started making flower beds. Since 2011, they have transplanted lots of flowers. They do this three times a year to contribute to the improvement of the community's landscape. They have also played an important role in VDPO activities. This will be discussed later in the paper. As Salon did not have its own base for communication, the community constructed an assembly house furnished with a kitchen and a washroom in 2012 for the group. It cost 1 million yen at the expense of the local government (500,000 yen) and the community (500,000 yen). The leader said that he recently felt "joy" in the community activities. For 2013, he plans to hold a residents' talent show in collaboration with neighboring Kainokawa-Gou community to share the feeling of "joy". He sees the stagnation of communal activities in the neighboring community. And since he knows well about the difficulty of community management, he considers the prospect of integrating both communities in the near future.

With respect to VDPO activity, the evacuation drill has been done every year since 2008. The residents' experience of "lobbying in the rain (2008)", which was mentioned before, triggered their desire to raise awareness of disaster risk reduction in the community. Some of the work they did is as follows:

- Volunteer Civil Work
  In 2009, they did maintenance and repair of evacuation stairs utilizing a local government subsidy (see Fig. 8.1). Since then, residents have engaged themselves in civil work for repairing evacuation routes. Traditionally they had a volunteer civil work system called as "deyaku". Since the olden days, all households have been required to participate in communal civil work. In 2012, they constructed a 25-meter-long evacuation route. According to the leader, such structural object has become a "symbol of community cohesion" for the residents.

Fig. 8.1 Repaired stairs

Fig. 8.2 Mowed evacuation route

- Discovery of Risk
  During the first evacuation drill (2008), elderly residents could not climb the evacuation stairs because there were many weeds. In the following year (2009), some residents voluntarily cut the weeds a few days before the drill (see Fig. 8.2). By clearing the paths, residents discovered that their evacuation routes and evacuation spaces were at risk of sediment disasters. Many of the routes and spaces had been utilized as terraced paddy fields and paths. As there are many abandoned fields, the residents have started discussions on necessary measures to be taken.
- Having opportunity to study disaster risk
  In August 2012, as the Cabinet Office officially announced the worst scenario of a Nankai earthquake, the community board members visited Nakanohama community (which promoted VDPO activities aggressively) to learn their progressive VDPO approach. Furthermore, they had periodical meetings with officials in charge of the VDPO to study disaster risk in the community (the activity was introduced in a city report in January 2013). Based on such meetings, they have considered possible measures against disaster risk. For example, they have devised a plan to utilize empty houses as places for storing water, blankets, and/

or something heavy ready to use in case of emergency. In addition, the leader has planned to implement an evacuation drill without advance notice at night in winter season to estimate worst case situation.

On 11th March 2012, the Great East Japan Earthquake occurred. During the earthquake, Japan Meteorological Agency issued a tsunami warning for 19 municipalities (including Tosashimizu city) in Kochi prefecture. Though the height of tsunami was 132 cm and a number of overturned fishing boats were reported in Tosashimizu city, there was no severe damage. At that time, Tosashimizu city government issued an evacuation advisory, and the VDPO warned the residents about the risk of tsunami. Nevertheless, the people stayed in their houses and did not evacuate to the designated area. Fortunately, the height of the tsunami was low (below 1 m). The designated evacuation site was the shrine, which did not have shelter. So, even if they could stay there temporarily, it would be difficult for the evacuees to spend a night. Moreover, because the roadway runs along the coastal line, evacuating by car would not lessen the danger posed by a tsunami. It is considered that such circumstances hindered the residents from evacuating. This experience highlighted the difficulty of evacuation during nighttime in winter.

8.4 Discussion and Analysis

8.4.1 Key Drivers to Start a VDPO Activity

The chronology of the activities in Kainokawa-ura community (Table 8.1) shows that it took about 5 years to implement the first evacuation drill since its inauguration. As mentioned earlier, at the establishment of VDPO, their communal activity stagnated. So, what drove the implementation of the first evacuation drill? Figure 8.3 is a comparison of networks between 2006 and 2008; it shows that there were two drivers, that is, the "development of a core communal group" and "collaboration with external entities". In 2006, there was no "core" communal group. Though the community had administrative staff, the leader practically fought alone, which made community management difficult. After the inauguration of residents' social welfare group, "Salon", its members, through the effort of the leader, came to understand the necessity of contributing to the activation of communal events such as summer festival. Gradually, they gained other residents' trust. After that, some members were elected as communal board members officially (following a communal rule), which led the group to play an important role in VDPO activities. With the assistance provided by the Salon members, it became easier for the administrative division as well as the VDPO to call for other residents to be engaged in communal activities. That is to say, the existence of Salon played an important role to connect communal organizations and residents. It worked as a foundation of communal management. The establishment of such a communal group met the potential need of the community. This means that the viewpoint of community development is indispensable for VDPO.

Table 8.1 The chronology of the community activities

		Communal events (except VDPO)	VDPO activities in the community	Issues regarding external entities
Stagnant stage	Before 2000	– Dissolution of a young generation's group (in 1980s) and a women's group (in 1990s)		
	2001	– *Flood damage*		
	2003			– Local gov't's (DRR section) proposal to establish a VDPO
	2005		– *Inauguration of VDPO*	
	2006			– Local gov't's (DRR section) proposal to collaborate with a neighboring community in its VDPO activity
Turning point	2007	– *Inauguration of "Salon". Salon members started supporting various communal activities (communal summer festival etc.)*	– Cancellation of the first evacuation drill	– *Start of discussion between the community leader and the neighboring community leader over VDPO* – *Local gov't's (social welfare section) support for the establishment of Salon*
	2008	– Assignment of three Salon members as board members of the community	– *Implementation of the first evacuation drill* – *Many residents lobbied for the improvement of evacuation stairs*	– *A neighboring community's support for the first evacuation drill*
Active stage	2009	– Issuance of communal newsletter to publicize activities of the organization – The birth of "Volunteer Group" (a new communal group)	– Maintenance and repair of evacuation stairs (done as volunteer civil works) – Continuation of the evacuation drill – Increase in residents' involvement for volunteer work (mowing) to reduce disaster risk	– Local gov't's (DRR section) Subsidy Allocation for VDPO
	2011	– Construction of an assembly house as a meeting space for Salon – Implementation of a talent show	– Continuation of the evacuation drill	– Local gov't's (social welfare section) Subsidy Allocation for Salon (construction of Salon's space)
	2012	– Introduction of Salon in the local gov't's public relation magazine	– Construction of the evacuation route (done as volunteer civil works) – Continuation of the evacuation drill – A study visit to a more advanced community with good VDPO program	– Advanced community's (Nakanohama community) support for VDPO

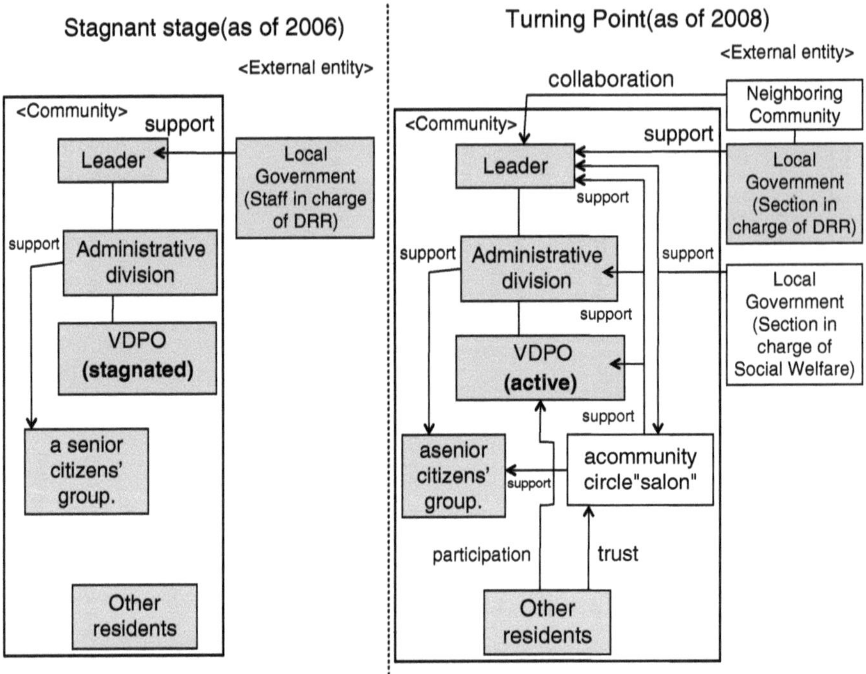

Fig. 8.3 Comparison of networks between 2006 and 2008

Under such circumstances, "collaboration with external entities" was an important factor. Without the support of the local government (social welfare section), the residents might not have an opportunity to develop such a core group. Moreover, through the support of the local government (disaster risk reduction section), the community was able to receive help from its neighboring community and implement the first evacuation drill successfully. Such external stimulation became the motivator to start the VDPO activity.

8.4.2 Key Factors to Enroot VDPO Activity in a Community

From 2008 up to the present, Kainokawa-ura community has continued its VDPO activities. Now the area that the VDPO takes up has expanded. Besides the evacuation drills, they have implemented study meetings with other advanced communities and members of the staff of the local government, and construction of evacuation routes utilizing traditional volunteer system (deyaku). Some changes have also been observed. The number of participants in VDPO has increased, and that some residents have come to mow voluntarily in the evacuation areas and routes so as to secure safer evacuation. It means that the residents' attitude towards VDPO has changed from being dependent on local government to being independent.

On the other hand, several studies (Takahashi 1995; Kuroda and Kagami 1998) have highlighted the stagnation of VDPO in many communities. Generally, the continuation of VDPO activities is not easy. What then is a key factor to promote VDPO activities? To provide an answer, we need to take a look at the effectiveness of social capital viewpoint. The concept of SC is currently receiving a lot of attention from development agencies and research institutions. In addition, many studies regarding SC have been conducted. SC is recognized as "the basics of governance in the economic society of the present age" (Miyagawa 2004). However, there are various definitions of SC which need consideration. It can be defined as "...features of social organization, such as trust, norms [or reciprocity], and networks [of civil engagement], that can improve the efficiency of society by facilitating coordinated actions" (Putnam et al. 1993). The World Bank (2000) has defined SC as "...the institutions, relationships, and norms that shape the quality and quantity of a society's social interactions". Narayan (1997) has defined SC as "...the rules, norms, obligations, reciprocity and trust embedded in social relations, social structures and society's institutional arrangements which enable its members to achieve their individual and community objectives". The analysis of Putnam et al. (1993) provoked controversy among many social scientists especially in the field of development. Studies on SC have proliferated since then, and the theory has been applied to different disciplines. Putnam (2000) focused on the function of SC and distinguished the difference between bonding and bridging. He stated that bonding relates to SC within a community of individuals who already know each other. On the other hand, bridging refers to connections among individuals who did not know each other previously. Bonding in SC is good for building specific reciprocity and mobilizing solidarity within a community. Bridging, by contrast, is better for building links to external assets and for information diffusion. As more detailed analysis has been conducted, several categorizations of social capital have emerged. Narayan (1999) indicated that the relationship between Bonding SC and Bridging SC is a trade-off. That is, if one community's Bonding SC increases, then Bridging SC decreases. Uphoff (1999) described a "structural" and "cognitive" SC. Structural SC involves various forms of social organization, including roles, rules, precedents and procedures, as well as a variety of networks that contribute to cooperation. Cognitive SC includes norms, values, attitudes, and beliefs. Structural and cognitive SC are complimentary: structures help translate norms and beliefs into well-coordinated goal-orientated behavior. Structural SC and cognitive SC are interactive and mutually reinforcing (Krishna and Uphoff 1999). It is widely agreed that SC facilitates mutually beneficial cooperative action, and it is defined in various ways. In this part, analysis is conducted from the viewpoints of four types of SC is shown: "bonding", "bridging", "structural" and "cognitive" SC. Hokkaido Prefecture (2005) considers that the four types of SC are correlates. It proposes a framework to examine public policy. Based on the framework of Hokkaido Prefecture (2005), the SC regarding VDPO activities in the community is pigeonholed (see Table 8.2).

Table 8.2 shows that the VDPO in the community has various structural/bonding SC. It also means that the VDPO has various interfaces with the residents.

Table 8.2 SC regarding VDPO

	Bonding SC	Bridging SC
Structural SC	– System for strengthening manpower utilizing communal resources (Salon members' support for VDPO) (Volunteer Civil Work System: "deyaku") – Information sharing system (issuance of communal news letter) – Systemization of VDPO activities (annual evacuation drill etc.)	– Support for implementation mechanism (in collaboration with the local government and with a neighboring community) – Learning risk opportunity (study meeting with an advanced community)
Cognitive SC	– Recognition of disaster risk in the community – Recognition of the necessity of preparedness – Recognition of other residents' needs for disaster preparedness – Feeling of cohesion in the community – Feeling of joy – Stimulation by other communal groups	– Recognition of the existence of advanced approaches at community level – Recognition of disaster risk in the city – Recognition of existence of bonds with the neighboring community

They utilize communal resources: the traditional system ("deyaku") and the social welfare group ("Salon") to strengthen the implementation mechanism of their organization. It should be noted that strengthening "Salon" shares a mutual interest for both the residents and the VDPO. Such approach allows VDPO activities to take root in the community. At the same time, the community utilizes structural/bridging SC as a driving force to promote VDPO activities. Though it is said that the relationship between Bonding SC and Bridging SC is a trade-off (Narayan 1999), the community effectively utilized the characteristics of both Bonding SC and Bridging SC.

On the other hand, the VDPO activities through various structural SC have enabled residents to experience abundant cognitive SC. As the Japanese government aimed to increase the number of VDPOs, the local government was forced to prioritize the inauguration rather than the cognitive aspect. The local government stressed the speedy establishment of the VDPO, giving directions to each community leader to prepare some documents to apply for permission to establish their VDPO after the example of VDPO manuals. In the case of Kainokawa-ura community, cognitive SC works as a driving force to promote further VDPO activities. It can be said that paying attention to accumulate cognitive SC is another key factor. Considering this, what aspect of cognitive SC should be paid attention to? According to the result of a survey done by Tosashimizu city government (Tosashimizu city 2012), for most communities in the city, joining VDPO activity is not "what they want to do" but "what they consider that they should do". That is, though they know the necessity of VDPO, they are reluctant to do so aggressively. It also implies that it is necessary to contrive a bridge between residents' disaster consciousness and feeling of "joy" through linking VDPO activities to other communal activities.

8.4.3 Necessary Support by the Local Government

In this chapter, only one case is analyzed through a field survey. In addition, controversy still surrounds social capital. Although it is difficult to generalize, the case of Kainokawa-ura shows that even a stagnant community has a possibility to enroot VDPO activities in the community. If a community is then stagnant, what should the local government pay attention to in order to promote VDPO activities? There are three necessary points to consider:

- Viewpoint of community development: The case of Kainokawa-ura community indicates that the viewpoint of community development is indispensible. Firstly, a local government is required to support a targeted community in clarifying internal and external resources, and issues in the community. Secondly, both the local government and the targeted community should consider possibilities to solve issues through VDPO activities. If necessary, it should prioritize the development of a "core" communal organization that might sustain communal management rather than hasty implementation of VDPO activities. In this aspect, the local government is required to have a cross-sectional discussion with the organization.
- Linking a targeted community with other external entity: In the case of Kainokawa-ura community, the local government played an important role in connecting the community with a neighboring community and an advanced community. That is, it contributed to expand bridging SC. Collaboration with a neighboring community brought the community leader confidence to implement the first evacuation drill. Through study meeting with the advanced community, the community learned possible approaches for disaster preparedness at community level. Moreover, the attitude of the community changed from local government-dependent to independent. Generally, a local government accumulates various knowledge and information regarding communities in the city. Local governments should promote mutual learning in order to share appropriate know-how and stimulate communities. As the number of local government staff is limited recently, such approach will support a local government's function complementarily.
- Monitoring changes of SC in each community: As the case of Kainokawa-ura community indicates, a "community" is like a creature. That is, SC can vary easily. The local government is expected to deliver information regarding changes of SC in each community. If the environment around the community changes, communal resources and residents' needs may also change. As Albee and Boyd (1997) argue, there is "*no single answer or model to promoting participation …*", local governments will have to reexamine their approaches in accordance with the change of SC in the society.

8.4.4 What Remains to Be Seen

The case of Kainokawa-ura poses a difficult question to us. That is, "what is the goal of community based disaster management?" Though the residents have "promoted"

VDPO activities, they did not evacuate during the Great East Japan Earthquake. To what extent are communities expected to behave when there is a risk of disaster? To what extent can local government act? How will we be able to evaluate their activity? It remains to be solved.

Ono (2005) indicated that the number of marginal villages, where residents (over 65years old) occupy over 50 % of the population, is increasing and it will be very difficult for these villages to maintain joint activities. Without joint activities, it will become more difficult for communities to take continuous countermeasures. In addition, with the number of marginal villages, the amount of abandoned farmland is also increasing. This neglect of the fields causes water shortage and flash flooding (Ono 1998). When the number of marginal villages increases (Ono 2005), their vulnerability to disaster increases. The importance of further studies on necessary countermeasures for disaster preparedness from the viewpoint of the community's aging rate will increase.

Eade (1997) stated, "*instead of passively waiting until their situation is more certain, people can come to see crisis as an opportunity to expand their horizons, envisage their own future, and take positive steps to realize their aspirations.*" It is impossible to halt the aging of a society that is undergoing depopulation. However, we should seriously consider the possibility of establishing a more resilient society to optimize the capacity of individual communities. The necessity of supporting our senior citizens has become more prominent in aging societies like Japan, and the solution requires an accumulation of experiences focusing on social capital for effective approaches.

Acknowledgements The authors acknowledge the support of the community and the local governments in conducting this study.

References

Albee A, Boyd G (1997) Doing it differently: networks of community development agents. Scottish Community Education Council, Edinburgh, pp 3–24

Cabinet Office (2007a) Bousaihakusho (White paper on disaster preparedness). http://www.bousai.go.jp/kaigirep/hakusho/h19/index.htm. Accessed 3 Feb 2013 (in Japanese)

Cabinet Office (2007b) Koureishakaihakusho (White paper on aging society). http://www8.cao.go.jp/kourei/whitepaper/w-2007/zenbun/19index.html. Accessed 3 Feb 2013 (in Japanese)

Cabinet Office (2008a) Bousaihakusho (White paper on disaster preparedness). http://www.bousai.go.jp/kaigirep/hakusho/h20/bousai2008/html/honbun/index.htm. Accessed 3 Feb 2013 (in Japanese)

Cabinet Office (2008b) Koureishakaihakusho (White paper on aging society). http://www8.cao.go.jp/kourei/whitepaper/w-2008/zenbun/20index.html. Accessed 3 Feb 2013 (in Japanese)

Eade D (1997) Capacity-building an approach to people-centered development. An Oxfam Publication, Oxford

Hokkaido Prefecture (2005) Social Capital no josei to chiikiryoku no kojo (Establishment of social capital and community empowerment). http://www.pref.hokkaido.lg.jp/ss/ssa/academy/17academy/sc/mokuji.htm. Accessed 3 Feb 2013 (in Japanese)

Imai Y (1968) Nihon no Kaso chitai (Marginal villages in Japan). Iwanamishoten, Tokyo (in Japanese)

Krishna A, Uphoff N (1999) Mapping and measuring social capital. http://siteresources.worldbank.org/INTSOCIALCAPITAL/Resources/Social-Capital-Initiative-Working-Paper-Series/SCI-WPS-13.pdf. Accessed 3 Feb 2013

Kuroda S, Kagami Y (1998) The present situation analysis of voluntary organizations for disaster prevention in Kawasaki City. Summaries of technical papers of Annual Meeting Architectural Institute of Japan, F-1, Urban planning, building economics and housing problems, vol 1998 (19980730), pp 853–854 (in Japanese)
Mimaki J, Takeuchi Y, Shaw R (2009) The role of community-based organization in the promotion of disaster preparedness at the community level: a case study of a coastal town in the Kochi Prefecture of the Shikoku Region, Japan. J Coast Conserv 13(4):207–215. doi:10.1007/s11852-009-0065-8 (Springer)
Miyagawa K (2004) Theory of social capital in social capital; Gendai Keizaishakai no Governance no Kiso (Social capital; the basics of governance in economic society of the percentage). Toyokeizaishinposha, Tokyo, pp 3–53 (in Japanese)
Nakamura K (2009) Evaluation of "Fureai-ikiiki Community Salon" as regional community. Jpn Soc Home Econ 60(1):25–37 (in Japanese)
Nara Prefecture (2012) Voluntary disaster preparedness organization. http://www.pref.nara.jp/22630.htm. Accessed 3 Feb 2013 (in Japanese)
Narayan D (1997) Voices of the poor: poverty and social capital in Tanzania. The World Bank, Washington D.C., pp 50–65
Narayan D (1999) Bonds and bridges: social capital and poverty. Policy Research Working Paper No. 2167. The World Bank, Washington, D.C.
Ono A (1998) Sansonkankyoshakaigakujosetsu (Environmental sociology in mountainous villages). Nousangyosonbunkakyokai, Tokyo (in Japanese)
Ono A (2005) The present state of less-favored areas and the direct payment system. In: Yoshizawa Y S et al (eds) Japanese less favored areas and regional revitalization. Tsukuba-Shobo, Tokyo, pp 51–101 (in Japanese)
Perry R, Lindbell M (2003) Preparedness for emergency response: guidelines for the emergency planning process. Disasters 27(4):336–350
Putnam D (2000) Bowling Alone: the collapse and revival of American community. Simon and Schuster, New York
Putnam D, Leonardi R, Nanetti R (1993) Making democracy work: civic traditions in modern Italy. Princeton University Press, Princeton
Shinohara S (1991) Kasochiiki no Henbo to Sanson no Doukou (Changes in under populated region and movement in mountainous areas). Taimeido, Tokyo (in Japanese)
Takahashi K (1995) A study on the present situations and problems of voluntary organizations for disaster prevention in Nagasaki City after the 1982 Nagasaki Flood. Shizensaigaikagaku (J Jpn Soc Nat Disasters Sci) 14(3):219–234 (in Japanese)
Tosashimizu city (2005) Prospect of the population and forestry (internal information) (in Japanese)
Tosashimizu city (2012) Result of the questionnaire survey on community development (internal information) (in Japanese)
Tosashimizu city (2013) Census as of March 2010. http://www.city.tosashimizu.kochi.jp/outline/jinkou.html. Accessed 3 Feb 2013 (in Japanese)
United Nations University Institute for Environment and Human Security (UNU-EHS), the Alliance Development Works/Bündnis Entwicklung Hilft and The Nature Conservancy (TNC) (2012) World Risk Report 2012. http://www.ehs.unu.edu/file/get/10487.pdf. Accessed 3 Feb 2013
Uphoff N (1999) Understanding social capital: learning from the analysis and experiences of participation, Social capital: a multifaceted perspective. World Bank, Washington, DC, pp 215–249
World Bank (2000) What is social capital. http://www.web.worldbank.org/WBSITE/EXTERNAL/TOPICS/EXTSOCIALDEVELOPMENT/EXTTSOCIALCAPITAL/0,,contentMDK:20185164~menuPK:418217~pagePK:148956~piPK:216618~theSitePK:401015,00.html. Accessed 3 Feb 2000
Yamazaki T (1994) Gendaino Juminsoshiki to Chiikijiti (Community organization and community governance in modern ages). Jichitaikenkyusha, Tokyo (in Japanese)

Chapter 9
Traditional Community-Based Disaster Management in World Heritage Site of Shirakawa Village

Chiho Ochiai

Abstract People have been fighting disasters such as earthquakes, floods, and fire to protect their lives, houses, and villages. Since most of the houses in Japan are made of wood, preventive measures against fire are prioritized. Historical and cultural houses and buildings often attract people's attention. However, more importantly, that cultural heritage can be understood to be the result of efforts by local residents who have been maintaining structures and handing down traditions and beliefs to protect these heritage sites over the generations. Shirakawa village, a UNESCO World Heritage Site, is located in an isolated, heavy-snow environment. It has a long history of protecting its traditional houses, which are made with thatched roofs that are vulnerable to fire. The people of this village have formed associations, such as the Residents' Association (RA) and Fire Volunteer (FV) force, and undertaken several activities to keep social order and prepare for disaster and emergency. For this study, an interview survey and participatory observation were conducted with RA and FV members. The study revealed that people in the village have been sustaining their traditional organizations and activities in spite of modernization and socio-environmental changes. This chapter focuses on the processes and mechanisms that have been established by local organizations and activities to protect the Gassho-zukuri houses, which are prone to fire. It will also examine how those mechanisms were developed and how they have helped to increase the awareness of local residents of the need to sustain the houses and the heritage of the local community. The analysis revealed that many activities have been conducted on a daily basis by several organizations as duty of community laborer, resulting in the establishment of a multi-layered structure. The study also found an interrelation between activities and awareness/spirits. Hence, establishing mechanisms that involve all local residents has been a key to maintaining the village's cultural heritage.

C. Ochiai (✉)
Graduate School of Global Environmental Studies, Kyoto University,
Sakyo-ku, Yoshida-Honmachi, 606-8501 Kyoto, Japan
e-mail: ochiai.chiho.2x@kyoto-u.ac.jp

R. Shaw (ed.), *Community Practices for Disaster Risk Reduction in Japan*,
Disaster Risk Reduction, DOI 10.1007/978-4-431-54246-9_9, © Springer Japan 2014

Keywords Community-based activities • Sustainability • Traditional • World heritage

9.1 Introduction

The attention given to community-based disaster management (CBDM) has grown lately due to the increasing number of natural disasters (Ministry of Interior 2010). Tanaka mentioned that past disaster studies mainly focused on the post-disaster stage; however, pre-disaster stage is equally important (Tanaka 2007). This pre-disaster study not only focuses on the disaster period but also understanding disaster within the context of people's daily lives.

Communities all around the world have always coped with natural hazards, such as fire, floods and earthquakes. In order to achieve sustainable CBDM, understanding and analyzing existing community preparedness mechanisms is essential. However, it has often been ignored (Tanaka 2005). Local mechanisms can be understood from a disaster management (DM) perspective by analyzing social organizations, local activities, human relationships, and local mentalities that have been established and reinforced on an ordinary, day-to-day basis.

Historical and cultural heritage often attracts people's attention, especially tangible assets. More importantly, cultural heritage has been maintained and handed down through generations by local residents who have advocated the awareness and recognition of its importance. In other words, cultural heritage is a result of these intangible efforts by every resident who have had to cope with disaster. Many studies have examined the conservation of buildings and landscape, but very few have focused on the social mechanisms developed to protect cultural assets in relation to CBDM.

All the rural villages have established community organizations, which are responsible for not only maintenance of social order but also disaster preparedness and emergency response in Japan. Eitaro Suzuki who is one of the founders of Japanese sociology, noted that Japanese rural community/society is "the social unification of all the cultural bases to function as the collective defence and the life cooperation" (Suzuki 1940). By Edo-period, basic social organizations such as a joint labor, a religious service, a fire fighting, and crime prevention were already established within the village (Torigoe 1985). To take a part in disaster preparedness and emergency was considered to be an obligation in the village life, and often young males were ones that are responsible (Torigoe 1985; Sakurai 1988). It means each village has a mindset of self-help and mutual-help to protect their own living and community from the disaster.

Since most of the houses are built of wood, preventive measures against fire have been taken in many different ways, both structurally and non-structurally. However, studies of Resident's Association (RA) and Fire Volunteer (FV) in relation to DM have been very limited and poorly, if at all, documented (Goto 2010). Understanding the traditional organizations and their sustainability mechanisms, which have been developed and maintained over many years, is worthwhile in studying and planning DM at the community level.

This chapter aims to identify the process and the traditional mechanisms of CBDM through considering ordinary daily life, which exists in the context of the history, culture and social systems that have protected a community's cultural heritage in a sustainable manner.

9.2 Study Methodology

The results of the study are based on a review of literature and fieldwork conducted between 2009 and 2010. This fieldwork consisted of semi-structured and structured interviews, participatory observation, and open and informal talks using micro-ethnography methods (Minoura 1999). The study was conducted by repeating these methods to understand the local mechanisms; findings were confirmed through daily conversations and events and observations were explained and interpreted through these processes. These processes also helped to establish a social relationship between local residents and the researcher and often helped the researcher secure honest opinions and understand feelings of residents not expressed during formal conversation.

The initial interview survey was conducted from February to March 2009 by interviewing 100 people who include district leaders and others were randomly selected in four districts[1] of the village. The residents were asked to identify or describe (1) their past disaster experiences, (2) their current activities and awareness of CBDM, and (3) the groups in charge of CBDM. The interview survey was conducted by visiting all the houses during the day time, in addition to making appointments with each district head and group, referred to as the "kumi" (explained in following section).

As a result of the initial survey, it became clear that two groups are involved in DM in the village, namely the RA and FV. Therefore, in the ensuing fieldwork, these groups were targeted by participatory observation of important activities conducted to understand their organizations and relationships among people by recording their conversation, following and observing their activities and participating in some activities. Several interview surveys were also conducted, targeting FV members' understanding of the social mechanisms and established processes along with the relationships among FV members, their roles in the community, and the mentality and awareness of being member of FV. The results of the interview surveys and observation are presented in following sections. The result of the structured survey, conducted in October 2010 explained in later section (see in Table 9.1).

[1] Shirakawa village has 16 districts. Four were selected based on their population and function; namely Iijima (agriculture), Hatogaya (public facilities), Ogi-machi (world heritage), and Hirase (hot spring inns) where most of population are concentrated.

Table 9.1 Overview of field survey (number) person

Date	Method	Target/number of interviewee	Contents/topic
February to March 2009	Interview survey (semi-structured), observation	Local residents (100 persons), RA	Disaster experiences, disaster related activity and awareness, groups in charge DM activity
June 2009	Interview survey (semi-structured) observation	FV and WF (34) FV	Activities and roles of FV and Women's fire volunteer group (WF) FV training/skill, fire patrol
October to November 2009	Observation	FV	Activities of autumn fire prevention week, activities of FV in local festival
	Interview survey (semi-structured)	Local residents, leader (10)	Activities of RA, maintenance activity for water supply system
June 2010	Observation	FV	Competition of FV training/skill
	Interview survey (semi-structured)	Fire Brigade (1) FV leaders (3)	Relationship between FV and the fire brigade, DM issues in village
October 2010	Interview survey (structured)	FV (28)	Reason and incentives for joining FV, Awareness of DM, Role of FV and relations with community

9.3 Characteristics of Shirakawa Village

Shirakawa village was selected as the study area because it is an area representative of traditionally rural Japanese villages that have maintained their traditional architecture, community heritage, and spirit. The village is located in the north-eastern part of the Gifu prefecture, where the living condition is often referred to as that of an "isolated island in the land" because of the heavy snowfall and austere mountainous environment, though this isolation was alleviated with the opening of an expressway several years ago (Editing Committee of Shirakawa village 1995).

One of the districts, called Ogi-machi in the village (Fig. 9.1), was constructed in the architectural style known as Gassho-zukuri (here after called "Gassho-house"). It is characterized by thatched and steeply slanting roofs resembling two hands joined in prayer (Fig. 9.2). The Ogi-machi, 45.6 ha in area, has a total population of 604 persons (149 families) (Shirakawa village 2009). Fifty-nine Gassho-houses have been preserved in the area, and it was selected as an Important Historic Buildings Preserved District of Japan in 1976, and registered as one of the historic villages of Shirakawa-go on the UNESCO World Heritage Site list in 1995. Figure 9.3 shows an overview of the Ogi-machi (Fig. 9.3).

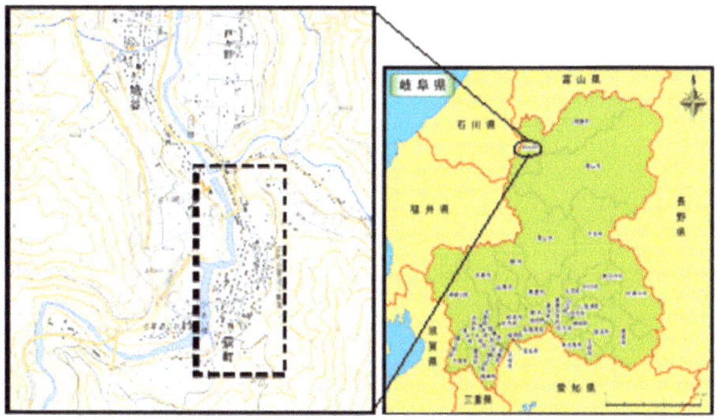

Fig. 9.1 Locations of Ogi-machi and Shirakawa Village (*Source*: Geospatial Information Authority of Japan, Web of Gifu Prefecture)

Fig. 9.2 Gassho House

Fig. 9.3 Ogi-machi

9.3.1 Community Organizations and Social System in the Village

9.3.1.1 Community Organizations in the Village

The village retains several traditional community organizations, formed along historical and thematic lines (Fig. 9.4). They conduct community activities to maintain the village lifestyle and prevent fire. Among these, the Resident's Association (RA)

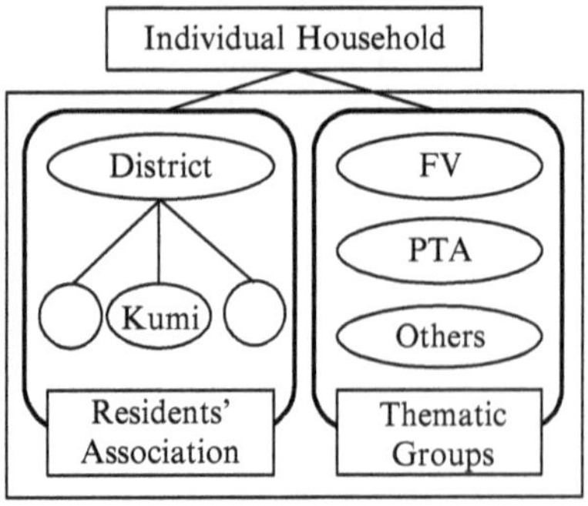

Fig. 9.4 Community organizations

is the most basic administrative unit; households are required to join and it is in charge of community management. The smallest administrative unit in each district is called "kumi," which is a location-based neighborhood group. Ogi-machi has eight kumi formed by about 15–30 households, depending on the area. There is a leader in each district and kumi, and the members of each kumi hold a monthly meeting to discuss local issues and convey messages from the higher-level administrative organizations (district to village).

On the other hand, the Fire Volunteer (FV) is a thematic organization formed specifically to cope with fire and disaster. There are many thematic groups in the village and residents join based on their occupation and interests, such as Parent–Teacher Association (PTA) and Commerce–Industry Association. Joining some thematic organizations such as the FV is almost obligatory but some are not.

9.3.1.2 Traditional Japanese Social System of "Yui" and "Ninsoku"

Japanese rural villages have been developing a traditional social system to work together in cooperation. There are several different forms depending on its requirements and names. Most common ones are "yui (labor exchange)", "moyai (work together)" or "tetsudai (assistance)" (Onda 2006).

The study area still maintains strong community ties in the village; they developed a labor exchange mechanism of "yui," and conduct community duties using community laborer known as "ninsoku." Yui is one of those systems seen particularly in agricultural communities where needs assisting one another through cooperative labor. For instance, rice planting which requires concentrated labor during a short period of time, the yui system was develop to leads people to work as a group to help one another to complete their work (The Board of Education of Shirakawa village 1989). In Shirakawa village, yui is used for replacement of the thatched roofs since it is huge and very steeply sloped, needs a tremendous effort. For one large house, about 200 people are gathered together to replace the thatched roofs in 2 days.

Another one is called "ninsoku" that is also a traditional social system used often in community activity requires the cooperation of each house to maintain the life in the village. It is conducted daily bases or periodical bases. For instance, ninsoku includes cleaning of agricultural canals, managing forest, stepping snow to maintain the road and others. Ninsoku is conducted as a duty of each household and fine is imposed often when they do not participate in the activity. These social system requires volunteer works are now recognized the importance to maintain the village life. The occasion to use this yui system has reduced but "yui-like" mindset, in other word, community bond still exist among residents as a cooperating and helping each other. Also, this community spirits of community bond is commonly shared among local residents and considered as the most important sprits that represent the village.

9.3.2 Gassho-Style Houses and Memories of Fire

The village has experienced many natural disasters such as floods and heavy snow. However, people are most afraid of fire because the houses have massive wood-frames and thatched gable roofs that are vulnerable to fire. Experiences and stories reinforcing this fear have been handed down over generations. The height of the house is about 8 m from the base to the roof top (Center for Ethnographical Visual Documentation 1995).

The Gassho-style house has a thatched gable roof with massive wood framed structure that supports the entire house. Bruno Taut, a well-known German architect reported his impression by "a roof similar to those seen in the medieval times in Europe". The second and third floors are located in the space created by the steep gable roof with a gradient of about 60° using rope and viscous neso, a type of rope made by twisting young Japanese witch hazel. The height is about 8 m from the bottom to the roof top (The Board of Education of Shirakawa village 1995).

Once the house catches fire, sparks fly up and spread the flames to nearby houses. There have been many incidents in which fires in the village burn several houses at a time. One of the biggest incidents in 1912 involved two neighboring districts near Ogi-machi; fire burned one temple and 57 houses simultaneously, which meant that about 70 % of all houses in the area were affected by this fire.[2] Many elderly people remember the incident occurred in 1956 it also burned six houses at once (Shirakawa village 1998). Both dire incidences occurred in neighboring district of Ogi-machi district. One said, "it was June 1956. Our teacher told us to go home soon after our lunch break at school. Many houses burned down quickly, and we did not have time to bring out furniture and house goods." Another said, "I was a member of the FV and rushed to the site. I was surprised to see how quickly the fire had spread. We splashed water onto the house from four corners, but the fire spouted from within the house, and we had to give up and run away. It was a scary experience and I have never been able to forget it."

[2] According to statistics from 1910, these two districts had 84 houses in total (see footnote 5).

In addition to this story, another said FV used to put a "red-mark" on house when fire risk was high, making residents anxious and encouraging them to give fire prevention undivided attention and maintain the neatness of their homes and yards. These stories remind the fear of fire and responsibility of each house that give influence to whole community, hence, to take proper action and conduct the responsibility/duty is important.

Consequently, people in the village are very conscious of fire, and each family has handed these experiences down by word of mouth. Even though the most recent large-scale fire occurred many years ago, younger generations know about it in detail. The stories remind people of their fear of fire and the responsibility each house has toward the whole community, hence reinforcing the importance of taking responsibility and proper precautions.

9.3.3 Activities of the Residents' Association

9.3.3.1 Fire Patrol Activity

The fire patrol, conducted by a community youth group, had been an important part of fire prevention activities throughout Japan in the past (Goto 2010). In Ogi-machi, people continue to conduct fire patrols every day to protect their houses from fire. Patrols are assumed to have begun by at the late 1890s for the purpose of issuing warning and raising awareness about the risks of fire among residents.[3] It is conducted four times a day, usually, at 11 AM, 6 PM, 8 PM, and 11 PM.

These activities are conducted by the ninsoku of the RA. The first three patrols are conducted at the kumi level, which is less strict and has an adjustable schedule that allows neighbors to take breaks. However, the last patrol is conducted at the district level and is mandatory. When one person finishes the patrol, he or she passes the tool on to the next household. Therefore, tasks are circulated independently, with each resident taking a random turn. For instance, at 8 PM, a person visits the houses belonging to his kumi one by one, dragging and making noise with the metal bar (Fig. 9.5). At 11 PM, the fire patrol is performed by a pair of neighboring houses. There are six stamps, located in different places within every corner of the district (Fig. 9.6). The patrol walks around for about 1 h to get all the stamps by making sounds with wooden bars. For this district level activity, residents must sign their names into a notebook that circulated among patrollers to confirm their activity. The summary of patrol is shown in Table 9.2.

Each activity uses different tools made from different materials with unique origins and meanings; some traditional tools relate closely to spiritual beliefs. A set of wooden bars called "hyoshigi" a traditional Japanese tool for raising

[3] When this activity started is not certain. The late 1980s were mentioned based on reference 11) and an interview with an elderly person. In addition, a wooden board, now replace by notebook, that formerly circulated is stored at Wada. The board is dated May 1923 and was made by the youth group.

Fig. 9.5 Traditional tools

Fig. 9.6 Patrol in district

Table 9.2 Summary of fire prevention patrol

	Time	RA group	Gender/ number	Method/tools	Characteristics and changes
1	11 AM	Kumi	Woman One person	Visit houses and advise residents to be "careful with fire"	Less often conducted because many woman are busy working at the tourism-related restaurant which are most busy
2	6 PM	Kumi	Woman One person	Wooden bar, walk around kumi area by making sounds	Conducted no more or less often due to social and lifestyle changes, lack of available people, and aging
3	8 PM	Kumi	Man One person	Metal bar Calling each house	Not conducted during winter because people cannot hear the sound of the metal bar
4	11 PM	District	Man Two person	Wooden bar, walk around district, take note	Conducted every day, mandatory in the district. Woman also conduct this patrol now

awareness of fire risk, and many people still recognize it by the sounds today. In the village, the wooden bars are believed to cut a line of fire made by the fire spirits who dance at night. Another tool used is a metal bar called "shakujo," held by the Shugensha, a type of holy man, when he walks for training, and it is unique in this region. According to local religious beliefs, the sound made by the metal bar has the power to cast a spell to purify evil spirits. During the winter, when they do not use these tools, the residents take good care of them. The tools have been used and repaired over many years and are considered somehow holy among people in Ogi-machi.

In an interview, one woman said, "[I] learned how to do the patrol and which houses to visit from my grandmother-in-law when I got married and came to the village. By going around the kumi area, she told me the character of each house; for instance, only one elderly person was living in this house, sleeping in the right end of the house." There is no written guide or manual for conducting the patrol, nor is there a record of which houses belong to which kumi. Knowledge of this is handed down by word of mouth within families to each subsequent generation, as is information about local residents and/or the neighbors. Some changes in fire patrol practices should be noted.[4] At the kumi level, fire patrol is conducted less in some kumi than in others. In some, only a few households conduct the patrol. One resident said, "[I] don't want to be blamed if something happens, so I do the task." This suggests that, in the eyes of neighbours, the activity is involuntary. The leader of one kumi said this issue was raised repeatedly in the monthly meetings and that certain people always opposed stopping the activity, in their generation and convincing others to respect the tradition and the importance of its continuity. Thus, it appears that the fire patrol is sustained by a combination of the extant community framework, neighbourhood relationships, reverence for traditional tools and spiritual beliefs, record-keeping systems, and consensus building. In addition, traditional practices are handed down by word of mouth within families through "learning by doing" and are supported by the traditional social system.

9.3.3.2 Patrol and Cleaning of Water Supply System

Ogi-machi has a water supply system (Kobayashi 1999) specially constructed for fire prevention,[5] which was installed following requests from local residents, who demanded giving priority to fire prevention for Gassho-houses. The RA adopted the responsibility of patrolling and cleaning this system as a ninsoku. The water storage tank is located at the top of a hill and the intake is upstream. Every day, one resident takes a turn checking and cleaning the water storage tank to prevent it from being blocked by leaves. The task itself takes about 1 h, and, in most cases, the head of the

[4] Several reasons could be listed, such as (1) there has been no large fire incident in these years, and younger generations and newcomers never experienced fire; (2) there are many women away from home during the patrol period because they work at tourism-related businesses; (3) younger generations or newcomers feel this is an irrational way to raise awareness; and (4) improved soundproof structures make it harder for people to hear.

[5] The water supply system includes 65 and 45 mm water hydrants (43 and 28 unites respectively), water gun (59 units), a gravity water storage tank (600 t), and water tanks of 40 t (8 units). This system was installed from 1976 to 1988. The duty of patrolling the system is entrusted to one person during the summer.

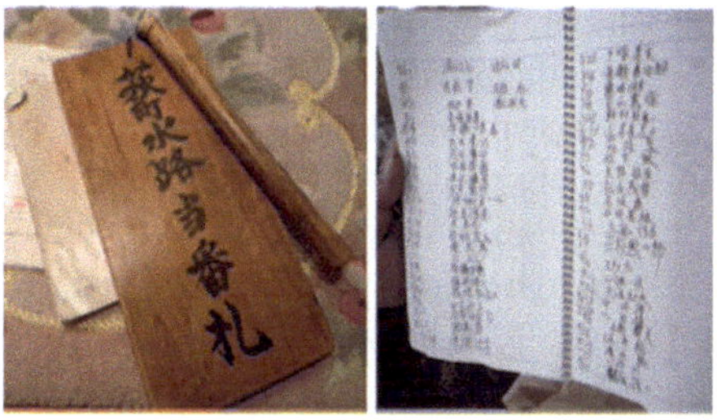

Fig. 9.7 Wooden board and notebook

Fig. 9.8 Activities of fire volunteer

family (a man or, in some cases, a woman, or two persons during heavy snow) performs the activity in the morning.

A wooden board and notebook circulate among families in the district (Fig. 9.7). When the board comes to a family, it means that the family takes on the patrol duty. When one family completes the cleaning, the head of the household writes his name in the notebook and passes it on to the next family. There is another notebook located at the water storage tank, and the resident writes his/her name there as well. This recording system confirms the date and allows clear identification of each resident's responsibilities. Since it is a ninsoku, this is mandatory and conducted daily. In addition to the patrol and cleaning of the water storage tank, several other activities must be conducted by the RA at the kumi, district, or neighbor level. For instance, one day during Fire Prevention Week in November, owners of Gassho-houses must practice spraying the water gun. Also, each household by taking tern is responsible to do snow plowing of water hydrant and water tanks constructed under the ground.

9.3.4 *The Fire Volunteer and Fire Prevention Activity*

9.3.4.1 Organization of Fire Volunteer

Fire Volunteer (FV) is the main organization to manage with any kind of disasters in the village (Fig. 9.8). The missions of FV are (1) to setting and maintaining the facilities, (2) to handle fire prevention, fire extinction, emergency response and rescue, handle in case of earthquake and strong wind and flood, and (3) to conduct disaster prevention and preparedness activities. The member of FV is appointed by the mayor and work as a part-time civil officer during the duty requested or permitted by the village. Basic salary and hourly wage in response to the deployed hours for duty are paid but it is considerably low. Also, all of FV has their job other than FV, therefore, it is considered as "volunteer".

Although Shirakawa village has had a fire brigade with 15 fire fighters working full time since 2000 but their mobility is limited, hence, even today, people in the village still have a great deal of responsibility in preventing and preparing for fires themselves. The FV[6] was organized in 1924 and has played an active role in the village since then. Ogi-machi was the first to establish an FV, and it later expanded to the whole village. There are 62 FVs registered in Ogi-machi in 2010, which means that about half of all households have one family member who is a member of FV. According to interviews, over 90 % of the men are members or former members of the FV.

The FV has over 50 events and activities yearly. These include routine activities, such as training and equipment maintenance. Major activities include practicing with fire-related equipment, searching for missing person, making sand-bag during heavy rain, snow-plowing, inspection of each house during disaster exercises week and etc. In addition to activities related to DM, the FV also manages community events, such as local festivals. Some activities are explained below. Figure 9.8 shows the FV in formal uniform and doing their duties.

9.3.4.2 Fire Prevention Inspection and Maintenance of Facilities

Twice a year, in spring and autumn, national Fire Prevention Week is held. The FV inspects all households one by one according to a 29-item regulation list unique to the village. The FV goes into each house, paying special attention to the kitchen area and the expiration dates of fire extinguishers and interviewing each resident. When the residents pass this check, the FV issues a certificate of inspection that has been prepared by the village for confirmation. This kind of activity is difficult to conduct because of privacy issues in the city; however, residents of the village collaborate to create a welcoming environment for FV. In addition to patrols by the RA, the FV clean the water storage tank during fire prevention weeks in spring and autumn. FV members go into the storage tank to clean out sediment during these periods. They also check for any cracks in the tank, and report these to the village office.

[6] In the village, 160 FVs are registered, and the FV is divided into three regional groups, one of which is in charge of Ogi-machi. The age of the FV members ranges from 18 to 50 years, with an average age of 34. The missions of FV are (1) construction and maintenance of facilities; (2) fire prevention, fire extinction, general emergency response and rescue, and natural disaster response; and (3) disaster prevention and preparedness. The FV members are appointed by the mayor and work as part-time civil officers during their terms of duty. All the FV members have jobs outside the FV. Basic salary and hourly wages are paid, but remuneration is considerably small. Therefore, it is considered a volunteer position.

Table 9.3 Reason for joining (multi. ans.)

Answer	Number (%)
Asked by Chief	22 (88)
Tacit understandings of villager	14 (56)
Saw as (almost) obligatory/mandatory	8 (32)
Father told to join	3 (12)
Did not know that one has to join	3 (12)

9.3.4.3 Roles in the Local Festival

The FV takes a major role in community activities not directly related to CBDM; however, many components of its involvement could be useful in disaster situations. The "doburoku" festival is held every October to appreciate the harvest and offer "doburoku," which means unrefined sake, to the shrine in hope of peaceful life in the village. This festival is the grandest of all the local festivities, and it determines the annual calendar. One of the festival's highlights is a parade, which has become over 150 m long and is joined by all residents regardless of age, who dress in traditional costume and walk around the village. Another highlight is the tasting of doburoku, held in the shrine. Over 10,000 people come to the festival over 2 days, and the small shrine fills with people.

When the village decided to highlight the parade to attract tourists, the FV took on the role of managing the festival. The FV adopted community activities as part of its duties according to the needs of community. During the festival, the FV plays a major role in managing the schedule and tourists, announcing programs, controlling traffic, preparing and carrying heavy equipment, preventing crimes and confusion, and carrying out related tasks. Many of these roles would be in emergency situations; therefore, the FV are implicitly training for disaster situations in many occasions of village life.

9.3.4.4 Roles of the Fire Volunteer and the Development of Volunteerism

The structured interview survey was conducted to clarify the FV's incentives, motivations, process of raising awareness, and roles in village organizations. Interviewees, selected based on their positions and ages, included 10 chiefs, 4 leaders, 11 members, and 3 retired FV. Ages ranged from the 20s to 60s. Many members joined the FV because a chief came to recruit them, and joining the FV is a tacit obligation in the village. One interviewee said, "Even though the FV is a 'voluntary' group, it is mandatory in this village." Another noted, "The day I came back to the village, [my] FV uniform was already prepared and it was left in front of the entrance [to my house]." Many FVs saw it as mandatory for men who decided to live in the village to support the FV system (see Table 9.3).

There is a national trend toward decreasing FV members in Japan (Yoshihara 2008). However, there has been no decrease of members of Ogi-machi since the 1970s. Some FV members shared positive views of their experience, one saying, "It

Table 9.4 Acquisition of FV sprits (multi. ans.)

Answer	Number (%)
Practice hard for the competition	11 (44)
Win the competition	8 (32)
Become a leader	8 (32)
Face actual disaster situation	8 (32)
Talk with other FV	6 (24)

is an honor to be asked to be a member of FV, because it means they recognize the person as a member of the community." There is no high school in the village, and young people usually leave at age 16. Some come back when they graduate high school, and some come back after several more years. Being an FV member gives people a chance to interact and create a close ties with other generations and families. Over half of all FV members mentioned their reasons for joining the FV included tradition, expectations, or family legacy. For many residents, protecting the village is common sense, deeply rooted in their mindsets. Therefore, they consider joining the FV to be their responsibility and part of village life.

The spirit and responsibilities of the FV developed over the years through activity participation, interaction with others, and education of young and new members. During the intense 2-week practice session for the FV competition in June, FVs spend many days and hours together. This not only improves requisite skills, but also builds human relationships and helps pass down traditions. Also, promotion to a higher position or the experience of actual disaster situations help members raise awareness and channel the FV spirit. When asked to explain their incentive to perform FV activities, FV members replied, "Because there are Gassho-houses there." FV members, like other local residents, know that Gassho-houses and the village in general must be protected from fire, and people are committed to this duty and the FV spirit (see Table 9.4).

The FV not only acts as a DM organization, but also as an organizer that is responsible for social activities and a place where young people who return to the village find acceptance, human connections, and relationships in the community. In the village as a whole, the FV has the respect of residents and plays major roles in all activities related to DM.

9.3.5 The Processes of Establishing the CBDM

Figure 9.9 shows the development process of CBDM in Ogi-machi. It shows how disaster experiences, historical events, social changes, and the development of the water supply systems influenced the development of the RA and FV and the establishment of certain activities.

First, CBDM activities began with a night patrol by youth group of RA before the 1890s, before being taken over by the RA. Although the night patrol was nationally encouraged, Ogi-machi expanded it and established a multi-layered mechanism according to their needs, and the continuity of the activity is outstanding. After the

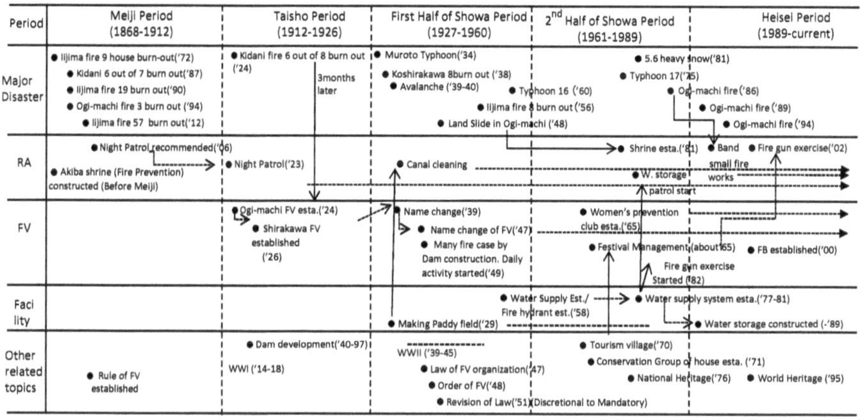

Fig. 9.9 Development process of CBDM in Ogi-machi

establishment of the FV in 1924, it played major roles in not only coping with disasters but also maintaining social order in the village.

As explained in parts of sections above, the RA and FV have adopted different activities related to DM and community activities from time to time, depending on their natural disaster and fire experience, village needs, newly constructed facilities, or the adoption of tools related to local religious beliefs and record-keeping systems. In some cases, the roles of each organization undergo changes, but the roles are still coordinated so that tasks are shared within organizations. There is a long process of developing the current CBDM. This process helped to mature the residents as a member of the community organizations, and also it helped those activities to be rooted in the community.

The process of establishing CBDM has mainly related to ensuring survival and protection of houses and preservation of villager's daily lives in an austere natural environment. Ogi-machi's status as a heritage site further increased awareness and incentives for residents to make a joint effort to conserve and protect the Gassho-houses. Therefore, residents of Ogi-machi have been incorporating natural and social environmental conditions and changes into their existing system and activities in a well-organized manner to ensure sustainability.

9.4 Discussion

This chapter has focused mainly on establishment and the sustainability of traditional disaster management specially fire prevention in the village. By examining the community organizations and local activities, several components can be pointed out how the community develops the awareness and voluntary spirits that is the key for organizations and activities to sustain. In this section, major elements and wisdom that discussed in regards to the existing mechanism and its sustainability.

9.4.1 Development of Community Organizations and Social System

People develop the mechanism according to live their life. Heavy snowfall and austere mountainous environment contributed to develop the social cooperation system of yui and ninsoku which are developed by the necessity and experiences, at the same time develop the "yui-spirit" or voluntary sprit. Also, FV has been maintaining the solidarity because there are Gassho-houses, where are their bases of their life, vulnerable to fire and danger of spreading fire.

Most of activities related to fire prevention are conducted by the local residents, under RA and FV. Each household owes responsibility of several activities as long as they live in the village and that contributes to keep the community to function as a whole. This is a fundamental components of typical "village community" where needs to maintain their life by themselves. These self- and mutual-help mindset and mechanism are driving forces behind community activity and duties of the community. Also, many told that "in order to protect their own house, community help is required, therefore, he/she also help others and participate in community work". This best represents the mentality and incentive of people in the area.

The residents conduct both activities by dividing their tasks into the different levels of local organization and alter their daily routine. Joining the RA and FV are mandatory, and also several daily activities are mandatory. In cooperating different organizations and different levels requirement in one activity create failsafe system. Although it gives a burden for residents, the "mandatory" component plays a key role to sustain the important activities in the village.

In addition, existing organizations adopted the new activity like the patrol and cleaning of water supply facility or management of local festival into their activities. This flexibility of organizations and task sharing established a multi-layer mechanism to have safety-net in every activity is valuable components of the structure of mechanism.

All of above mentioned components are based on the decisions made in the monthly and yearly meetings that all the residents are required to participate. Community organizations are open to a new idea or opinion and the democratic meeting system is in a way flexible to social changes but at the same time, it plays key role to enforce the rules and sustain the traditional activities.

9.4.2 Hand Down the Tradition and Increasing the Awareness

As explained in the beginning, the fire experiences are hand down from generation to generation by word of mouth which is subconsciously built the awareness toward fire and community ties of local residents. Most of community activities do not have written guide or manual for "how to" conduct these activity. Activities are handed down by participating in the activity as member of family or community.

Also most of the activities are conducted by the residents, both women and men. Since all activities are shared among the residents within a fixed framework, all are able to gain and share their knowledge, and understand the importance of their and the others' activities. Also, the participation of all residents gives a common topic of discussion and the ability to share knowledge, which contributes to increased awareness of the residents for disaster preparedness and it results the sustainability of activities.

In addition, such as fire patrol, the system of circulating and confirming the activity is simple and unique by using different tools. The wooden bars, iron bar, wooden board and notebook are used for circulating the activity and, recording and confirming the activities. The meaning of sounds is a shared knowledge and conducting the activity increases the awareness of residents. In the village life, spirituality has respected and believed for many years. Some of tools are rooted in spiritual believes. The activities combined with the spiritual component are continued for many years. It is not rational in modern society to believe in the sprits but this unseen believes somehow gives respects and becomes negative driving force not to stop the activity.

9.4.3 Mechanism for Sustainable Fire Prevention

The mechanism for sustainable fire prevention is developed by combinations of environment and historical experiences that created a system of organizational structure and activities that is well structured. Many activities related to fire prevention are conducted on a daily bases by local residents under the different community organization and often "mandatory". It established a multi-layered structure that creates the system of failsafe.

There are interrelations among local organizations, activities and awareness. Local residents developed their awareness and voluntary spirits by participating joining the community organization and conducting activities. Handing down their knowledge and wisdoms to next generation by word of mouth is a key component that is based on the human relationships in each house and also in the community. Also, people in the village inherit the cooperative spirit of "yui" and "ninsoku," including not only mutual but also self help, and this makes a fundamental and shared spirits among local residents. Figure 9.10 shows the mechanisms of CBDM in the village. It shows the relationship between community organizations and activities and, in connection to this, how history, nature, and environment influenced the mechanisms to develop social ties and trust, awareness, and responsibility sharing.

Although intensive participation of local residents in several activities are observed in Shirakawa village, these social organizations and their mechanisms are not unique to this village but also seen in other villages and town throughout Japan. Development of social organizations which encourage the development of social collectiveness and voluntary spirits with social responsibility, in other words obligation, are the key to sustain the CBDM.

Fig. 9.10 Mechanisms of CBDM

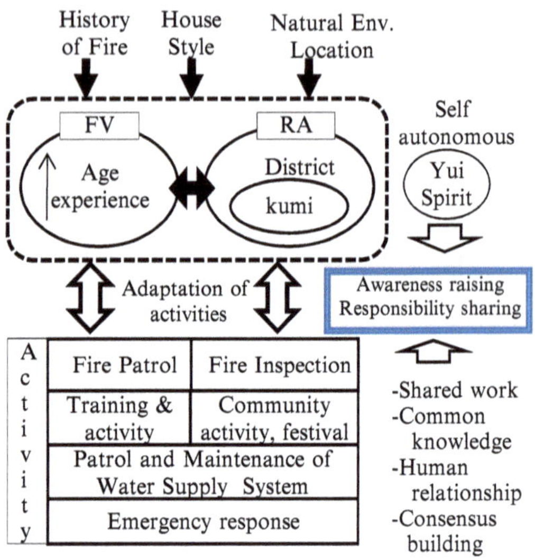

9.5 Conclusion

Local residents developed a CBDM mechanism that shares community duties among residents within a fixed framework, structured by multi-layered organizations and social systems so that activities have been conducted at regular intervals. Isolated austere environment, along with necessity and experience, has contributed to the development of the community laborer and social cooperation system of ninsoku, help to preserve the Gassho-houses and village community.

Local residents are conducting so many activities including some of laborer and undertake many roles in their everyday life. Sustaining the community is possible only with many regulations and limitations of the life, and that is why their traditional wooden houses as a traditional district are protected and inherited. Local residents in this village have been accepting these rules and regulation to their daily life. This mechanism developed over many hours and years spent working together and developing a trust relationship, which is a strong base for the CBDM to be sustained.

At the same time, voluntary spirits and responsibility to the community are developed through interacting with people in the community by joining the community organizations, conducting activities together and listening to the stories and experiences from older generations. Fire prevention is one of all the other responsibilities that local residents share. It is rational to accept the occurrence of fire and prepare it with effort of whole community. People in Shirakawa village were able to establish the sustainable mechanism for fire prevention; therefore, the cultural heritage of traditional houses and village life has been preserved and will be preserved in the future together with the mechanism for fire prevention.

Finally, all these finding are based on the field survey interviewing people. It is important to understand and analyze the existing traditional mechanism that is developed over the years in the village. Although each case must be unique, some components could be shared with other cases in different countries. The cultural heritage has been inherited because there are many efforts of local residents and there are always local wisdoms behind. To re-evaluating and re-recognizing the existing mechanism is the key for sustainable disaster management.

Acknowledgements The support and cooperation of local residents, fire volunteers, and local government is highly appreciated.

References

Center for Ethnographical Visual Documentation (1995) The construction of Gassho-style houses, 7 pp (in Japanese)
Goto I (2010) Shouboudan no Genryu wo Tadoru (Tracing back the roots of fire volunteer). Kindai-Shou ou-sha (in Japanese)
Kobayashi M (1999) Water supply system rooted in culture and climate for earthquake fire. J Seismological Soc Jpn (Second Series) 52:199–212 (in Japanese)
Ministry of Interior (2010) Bousai-hakusho, heisei 21nendo (Disaster prevention white paper, 2010 edn) (in Japanese)
Minoura Y (1999) Field work no Gihou to Jissai (Skills and facts of field work). Minervashobo (in Japanese)
Onda M (2006) Sougo Shakai Ron (Sociology of mutual help). Sekai Shisou sha (in Japanese)
Sakurai T (1988) Kou shudan no kenkyu (Study of Kou). Yoshikawa koubun-kan (in Japanese)
Shirakawa Village website, population of April, 2009 http://shirakawa-go.org/government/. Accessed Mar 2010 (in Japanese)
Shirakawamura kyouiku iinkai (The Board of Education, Shiwakawa village) (1989) "Yui" to Gassho, Shirakawamura ogimachi dentoukenzoubutugunchiku (in Japanese)
Shirakawamura kyouiku Iinkai (The Board of Education, Shiwakawa village) (1995) The construction of Gassho-style houses, a record of traditional techniques in Shirakawa-go (in Japanese)
Shirakawamurashi hensan iinnkai (Editing Committee of Shiwakawa village) (1995) Shinhen Shirakawa sonshi (History of Shirakawa Village) (in Japanese)
Suzuki E (1940) Nihon nouson shakai genre (Fundamental of Japanese rural society). Mirai-sha (in Japanese)
Tanaka J (2005) Saigai to Shakai, Heisei 17-nendo Houkoku (Disaster and society, 2005 year report) (in Japanese)
Tanaka S (2007) Kyoudousei no Shakai-gaku (Collectivity of regional sociology). Harvest-sha (in Japanese)
Torigoe H (1985) Ie to mura no shakaigaku (Sociology of family and village). Sekaishisou-sha (in Japanese)
Yoshihara N (2008) Bousai no Shakaigaku Bousai Community no Shakaisekkei ni Mukete (Sociology of disaster prevention and management). Toshin-do (in Japanese)

Chapter 10
Community Practices of Transmitting Daily and Disaster Information in Reihoku Area

Kumiko Fujita and Rajib Shaw

Abstract There are two major measures for disaster risk reduction, structural and non-structural measures. Early warning and evacuation are the major non-structural measures which are based on judgment and action of local people. Their judgment and action are decided by information through various media. In this paper, relation of media and local people in Reihoku area is analyzed. Reihoku area, Kochi prefecture is a landslide disaster-prone area. Possible information system for disaster risk reduction in Reihoku area is discussed based on the result of a questionnaire survey.

Keywords Community practices • Disaster risk reduction • Landslide • Transmitting information

10.1 Disaster and Information System

Japan is vulnerable to various disasters because of weather conditions such as typhoons and localized torrential rain, and topographical conditions. The mountainous and hilly areas make up 70 % of the total land, and a large number of rapid rivers and its geological weakness make Japan vulnerable to various disasters. Especially landslide disaster is a common disaster in mountain. Moreover, since Japan has already developed urban cities, urbanization has caused different type of landslide disasters. Occurrence of localized torrential rain is increasing, and big scale local floods and droughts will be occurred (Meteorological Agency 2007).

K. Fujita
UNESCO, Jakarta, Indonesia

R. Shaw (✉)
Graduate School of Global Environmental Studies, Kyoto University,
Sakyo-ku, Yoshida-Honmachi, 606-8501 Kyoto, Japan
e-mail: shaw.rajib.5u@kyoto-u.ac.jp

R. Shaw (ed.), *Community Practices for Disaster Risk Reduction in Japan*, Disaster Risk Reduction, DOI 10.1007/978-4-431-54246-9_10, © Springer Japan 2014

In addition to the climate change, social changes cause different types and scales of damage. Aging issue is one of them. Especially aging in mountain villages is more promoted than aging in urban cities.

There are mainly two measures against landslide disaster that are structural measures such as physical construction to reduce or avoid possible impacts of hazard, and non-structural measures such as policies, awareness and knowledge development. Structural measures have been implemented in major dangerous areas. Structural measures such as dam and embankment construction are not enough for responding to all landslide disasters caused by the increasing number of occurrence of localized torrential rains. Therefore comprehensive approaches are needed. Importance of non-structural measures has recognized in early 1990s. Issues of non-structural measures are in land-use practices, promotion of designating landslide disaster hazard areas, dissemination of evacuation and warning before disaster occurrence, security of evacuation place and route, and people's awareness of landslide disasters. Evacuation is the most closely depended on the judgment and action of local people. Local people decide whether they evacuate or not depended on their experience and disaster information. Social media such as Twitter, Flickr, Facebook, and blogs, and videos posted on YouTube has recently played a critical role in natural disasters as an information propagator that can be leveraged for disaster relief (Gao et al. 2011). There are two specific information dissemiantion systems which get attention in Japan in recent days. One of them is community radio, and the other is modern information technology, through social media.

Community FM broadcasting in Japan started with the 'Teletopia plan' (telecommunication and utopia) for promoting local activities advocated by the Ministry of Posts and Telecommunications (MPT, predecessor of the present Ministry of Internal Affairs and Communications, MIC) in 1983 (MIC 1985; Kanayama 2007). MPT published a report entitled 'Meeting for broadcasting in new media era' in April, 1987. MPT reported that they had to consider possibilities of introducing smaller FM stations to address local people's information needs (MIC 1987). Consequently, MPT announced the new establishment of community broadcasting targeting for some parts of municipalities in July, 1991. In 1992, community broadcasting was regulated. In the first 4 years after the first community FM broadcasting station started (from 1992 to 1995), the number of broadcasters increased gradually. In 1995, just after the Great Hanshin-Awaji Earthquake, the power limitation was pulled up to 10 W so that broadcast can be listened around a 5 to 10 km radius. Community FM broadcasting began getting known in the society. Starting from the very next year 1996 until 1998, a great many broadcast stations were opened around Japan. However, in the next 5 years (from 1999 to 2003), poorly managed five stations were closed and the opening of new stations slowed down. But after 2004, due to many natural disasters like mid-size earthquakes, typhoons, and heavy rains, the number of stations launched began increasing again. During this period, different kinds of NPO stations have appeared. Kyoto Sanjo Radio Café, founded in 2003 and the first NPO station, triggered a new movement in Japan. The reasons for the continuous upward trend are the relaxation of power limitation by the government, the opening of community FM broadcast stations, and the re-realization of the importance of community FM (Ideta 2012).

The use of social media by individuals and organizations differs based on their needs and situation during disaster. Individuals are able to use social media during disaster to receive vital disaster information related to safety, shelter, people finding, and basic necessities. Individuals are able to receive this information as well as provide it for others who are in need of these resources. This fact has led to the ability to provide informational support to information seekers through crowdsourced collaboration through cooperating with other like-minded individuals. In many cases of such collaboration, the groups are not ones which existed prior to the disaster, but spontaneously created groups of individuals who "just wanted to do something," demonstrating the potential for non-location specific volunteering during disaster. Through this collective gradual discovery of the possibility of social media in disaster, the evolution of social media's use in disaster, innovative disaster-specific social media types have also been developed, such as the crisis mapping system Ushahidi (Peary 2012). Due to the likelihood that social media use will grow around the world, it is also likely that there will be increased use of social media in disasters in coming years. In order to address problems related to social media use and better respond to disasters in which a greater portion of the population turns to social media or relies on it exclusively, institutionalization of social media use in disasters is necessary. Through understanding the situation regarding the use of social media in disasters organizations and government bodies may help to provide more reliable information-based support for affected individuals and improve existing methods of support by directing individuals to organizations where they may be helped or be of help.

In this chapter, daily and disaster information delivering system and propagator in Reihhoku area is analyzed based on questionnaire survey. Then possible information system for disaster risk reduction in Reihoku area is discussed.

10.2 Characteristics of Study Area

Reihoku area, Kochi prefecture is located in the center of Shikoku Island as Fig. 10.1 shows. The area is about 757 km^2 with three towns (Otoyo, Motoyama, Tosa) and one village (Okawa). Major road is national route 439 which across Reihoku area. Most of the public facilities, big hospitals and shops are located along the national route 439 (Reihoku Administrative Office 2009). Administrative organizations in Reihoku area are concentrated in the center of Motoyama town. The biggest grocery store and the biggest general hospital in Reihoku area arc in Tosa town.

Reihoku area locates in the upper Yoshino river basin. Landslides in Yoshino river basin are one of the largest landslides in fracture zone (Otahara 1991). Sameura Dam locates across Motoyama town and Tosa town in the upper Yoshino river basin, which is called water jar of Shikoku Island. It is mountainous with steep mountains of 200 to 1,700 m high in the north and forest shares 87.9 % of Reihoku (Table 10.1). Agricultural land shares 1.7 % and residential area shares only 0.6 %, therefore it is a typical mountain village (Reihoku Administrative Office 2009). In addition to the steep topography and fragile geology, the area has been suffered from heavy rain frequently because of the effects of moist airstream from the Pacific Ocean (Kochi Prefecture 2006).

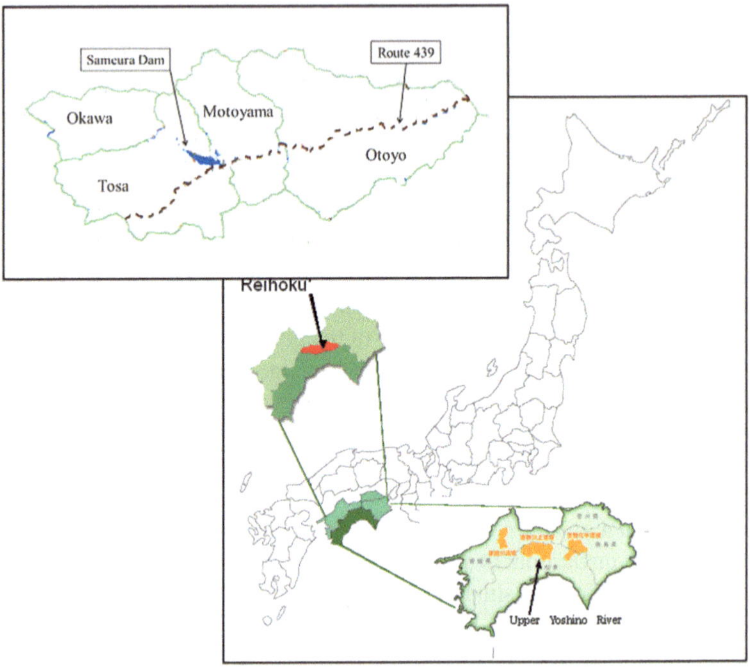

Fig. 10.1 Reihoku area, Kochi, and upper Yoshino river basin, arranged from (*Source*: Shikoku Regional Development Bureau 2010)

Table 10.1 Land use of Reihoku Area in 2002 (Reihoku Administrative Office 2009)

Category (ha)	Farmland	Forest	Water	Road	Residential area	Others	Total area
Otoyo town	363	27,654	470	572	163	693	29,915
Motoyama town	312	11,912	208	208	88	2,272	15,000
Tosa town	523	18,127	786	339	102	1,334	21,211
Okawa village	58	8,809	177	124	13	347	9,528
Total	1,256	66,502	1,641	1,243	366	4,646	75,654
Ratio in Reihoku	1.7 %	87.9 %	2.2 %	1.6 %	0.6 %	6.1 %	100.0 %

10.2.1 Major Disasters and Trend of Recent Disasters

There were four major sediment-related disasters in 1970s. Since the area had been hit by heavy typhoon storm in 1975 and 1976 and many mountain collapses were occurred, the area had devastated and chronic muddy water in Sameura dam became a social issue. Then national government has been managed the upper Yoshino River basin directly for preventing mountain disasters and related-issues since 1979 (Kochi Prefecture and MLIT 2004). Since then, there was no major disaster until 2004.

Okawa village and Tosa town were hit by abnormal torrential rain in August 2004. During that year Kochi prefecture was hit by typhoons and localized torrential rains. It is the record disaster year that a record number of five typhoons hit there (Kochi Newspaper 2004). It is so-called Sameura torrential rain. The accumulative precipitation of Sameura torrential rain is more than 1,000 mm in 3 days, which is more than one third of the annual precipitation. Because of the high precipitation, mountain collapses were occurred mostly in the east part of Okawa village, and the roads were cut off, as a result, whole Okawa village and Minami-kawa settlement in Tosa town were isolated, however, there was not casualty. One of the villagers in Okawa village recognized the abnormal change and contacted village town. Since the villager also called for evacuation to neighbors, there was no human damage. The damage was minimized, because awareness of disaster prevention and mutual cooperation among residents is strong in Okawa village (Kochi Prefecture and MLIT 2004). Mayors of Okawa village and Tosa town gave suggestions to improve disaster risk reduction system. Mayor of Okawa village emphasized that governments need to take responsibility of providing and maintaining structural measures, and self-, mutual- and public-help is needed for saving lives by ourselves, in addition to maintaining disaster prevention equipment (Gouda 2007). The mayor of Tosa town also emphasized the necessity of self-help organization. In addition, establishing information collection system is needed, especially in case that settlement becomes isolated (Nishimura 2005). Issues of helping old people to evacuate are also mentioned after Sameura torrential rain.

10.2.2 Social Changes

The rate of 65-year-old is increasing and the number of people is decreasing in Reihoku area (Fig. 10.2). The total population has been decreasing from 34,801 in 1965 to 15,270 in 2000 and the ratio of more than 65-year-old has been increasing from 11.6 % in 1965 to 38.7 % in 2000. It is estimated that the population decreases to 12,623 and the ratio of more than 65-year-old increases to 44.7 % in 2015. In addition, the number of people who are certified by the municipal government as being in need of care is tend to be increased, because the rate accounts for 12 % of total Reihoku population 16,786 in 2001 and increases by 15 % of 15,815 in 2004 (Reihoku Administrative Office 2009).

Spatial distribution of aged population (Kochi Prefecture 2010) shows that less aged population gathers in the center of Motoyama town with administrative organizations and along the national route 439 with big hospitals, schools and big grocery stores. Thus younger generation tend to live near major roads where public facilities, hospitals and grocery stores locate, and older generation tend to live far from major roads, public facilities and shops. In addition, the population in each community, especially in mountains far from route 439 is smaller and each community is located separately.

Fig. 10.2 Demographic change in Reihoku area (Reihoku Administrative Office 2009)

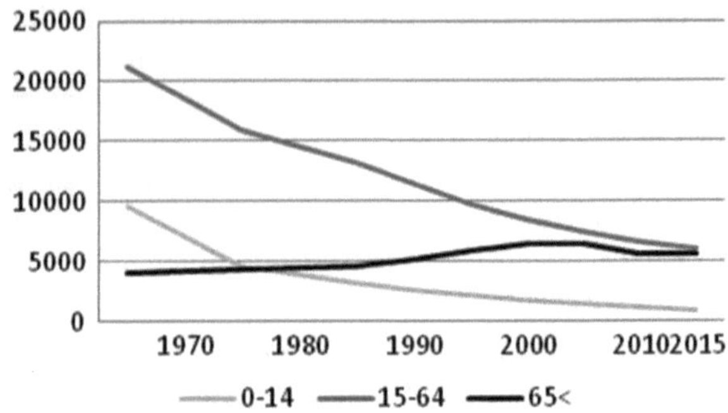

10.3 Questionnaire Survey in Reihoku

A questionnaire survey was implemented to Reihoku local residents with the cooperation of Reihoku NPO. The purpose is to know the people's practices and how it is related to disaster risk reduction. Three hundred and forty-seven people were selected from 5,267 registered lists of the phone book at random. Questionnaires and return-mail envelopes were sent to the selected people in December, 2009. The registered numbers on the phone book, selected numbers in each town and village are in Table 10.2. Since a supporter of Reihoku NPO who is from Okawa village had wishes to send questionnaires to all Okawa people, the sets of questionnaire survey were sent to all Okawa villagers on the phone book. Seventeen were returned, mainly because they moved out. One hundred and fifty-five people replied until the end of 2009. Eleven of the 155 reply are not available, mainly because they returned just to inform that the selected people were died. Then the collection rate is 45 %.

One hundred and forty-four replies are analyzed in this research. As the basic information of the respondents, 79 % (113 people) are male, and 17 % (25 people) are female. Six people had no answers about gender question. Figure 10.3 shows age range in each town and village. As the percentage in all Reihoku area shows, there are few respondents who are younger than 49 years, and more than 60 years old shares almost three forth. Tosa town shows higher rate in the age range of 70s and Okawa village shows higher rate in the age range of more than 80 compared to others.

10.3.1 Information and Communication Tools

It is important to know how people obtain daily information and disaster information, and how they deliver information each other. The result will suggest the possible communication and issues in case of disaster. As Fig. 10.4 shows, both daily and disaster information is obtained by TV and Kochi local newspaper.

Table 10.2 Number of distribution and reply

Town/village	Number of people on the phone book	Number of selected people	Number of respondent (male, female, no answer)
Otoyo	2,311	54	21
Motoyama	1,325	54	20
Tosa	1,444	52	25
Okawa	187	187	78
Total	5,267	347	144 [male: 113 (79 %), female: 25 (17 %), n: 6]

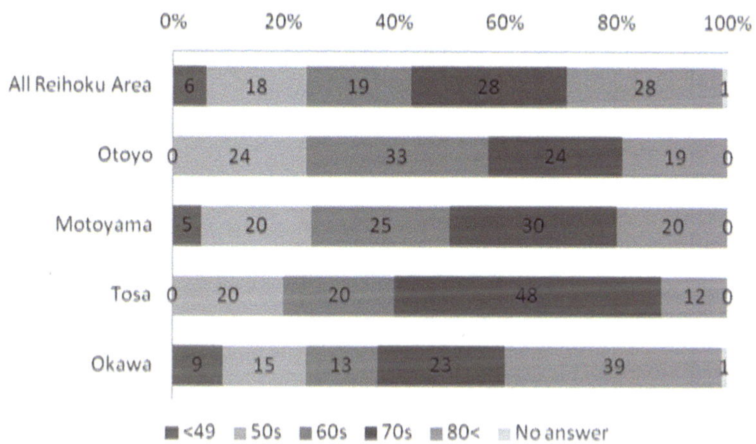

Fig. 10.3 Percentage in each district by age range

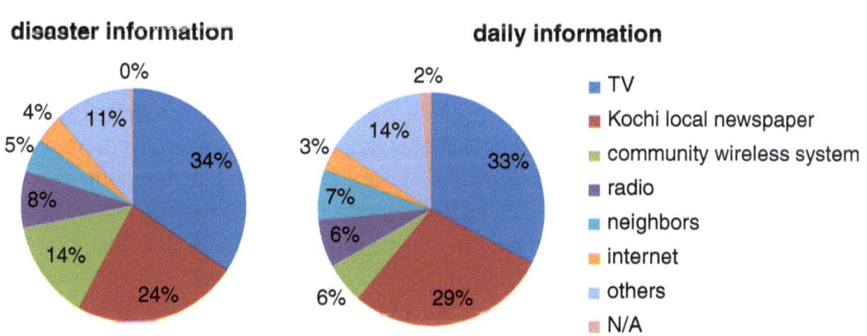

Fig. 10.4 How people get disaster information and daily information

There are three major characteristics of TV as a media: prompt report, simultaneous broadcasting and reliability. Shift from analog broadcasting to terrestrial digital broadcasting has been promoted and completed in 2011 as a national policy, because terrestrial digital broadcasting has characteristics of record performance and

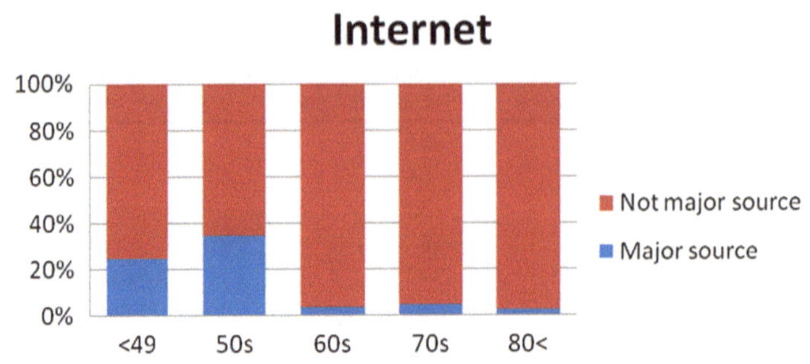

Fig. 10.5 Internets as a major disaster information source for younger generation

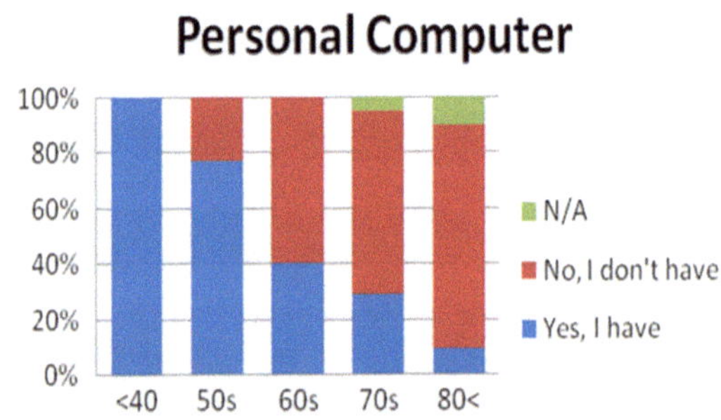

Fig. 10.6 Respondents who have PC (*left*)

versatility to other digital media in addition to three major characteristics of TV (Isono 2009). Broadcasting disaster information by terrestrial digital broadcasting is superior to other existing media in all points of speed, simultaneous broadcasting and accuracy. Since it is proved that it is an epoch making IT tool, broadcasting disaster information is expected (Isono 2009).

In case of disasters, the rate of listening community wireless system and radio is increased. Most of the comments are quite favorable to community wireless system. Community wireless system used in Reihoku area is called off talk telecommunication service. It is a local information delivery service through telephone line of analog fixed-line phone and ISDN. Terminal adapter is installed in each house. The delivered information is not limited only emergency information like disasters, but local daily information. Since off talk telecommunication service is a wireless system, information is delivered even in case of black out. Tosa town and Motoyama town promote to shift the system from off talk telecommunication service to IP message broadcast system because the terminal adapter has been used for a long time.

Many respondents, young or old, answered TV, local news paper and community wireless system is the major information source during disasters. However, younger generation answered internet is also a major source for getting disaster information (Fig. 10.5).

The reason is analyzed by possessing rate of personal computer by age range (Fig. 10.6). There are a few differences in the use of the function of personal computer.

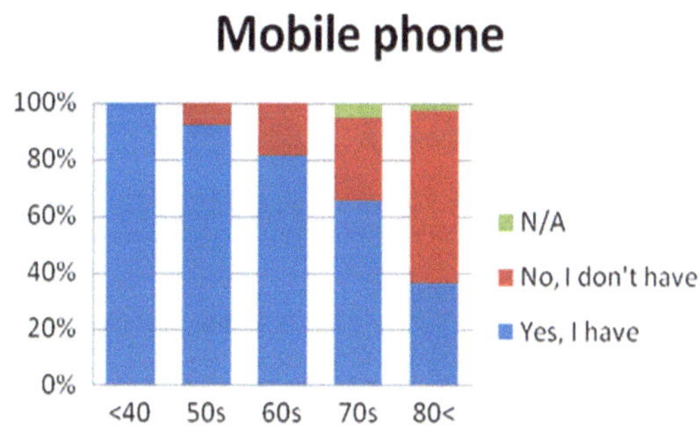

Fig. 10.7 Respondents who have mobile phone (*right*)

Table 10.3 Difference of receiving and sending disaster information by young and old people

	Young people	Old people
Receiving information through	TV Newspaper Community wireless network Internet	TV Newspaper Community wireless network
Receiving information from	Various media	Less various media
Sending information through	Email and internet function of PC and mobile phone	Talking Phone
Sending information to	Many people (including unknown people)	A well known person

Older generation use email function mainly. However, younger generation use various functions such as excel, ward, blog and internet search. In addition, younger generation use personal computers frequently. Similar trend is seen in the possessing rate and the use of mobile phone. However, the possessing rate of mobile phone is more than 60 % even in 70s (Fig. 10.7).

Similarly the function of mobile phone used by older generation is simple like only telephone function. Younger generation use more various function of mobile phone such as email and internet. Thus it can be said that younger generation have more possibility to gain public information through various media and to send information to various people at once (Table 10.3). Old generation is more passive about disaster information. Young generation was also rather passive about disaster information before mobile phone and internet become popular, especially before the function of blog, twitter and Facebook become popular.

10.3.2 Networking with Neighbors

Besides the system of media, there are issues on transmitting DRR information in human communication: "transmitting information to local residents", "interface issues" and "lack of communication by the person in charge." Especially human

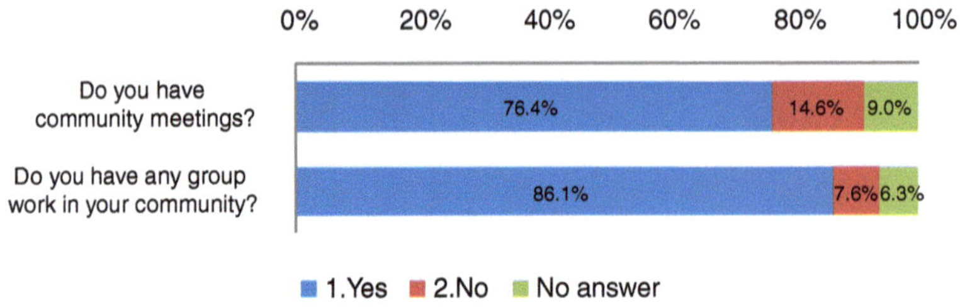

Fig. 10.8 Group/community-based communication

Fig. 10.9 Individual communication

communication is the key and basis of the transmitting information. Since communication with neighbors is also important in the point of promoting mutual help, it is important to know the community communication in Reihoku area.

Questions asked were "Do you have community meetings?", "Do you have any group work in your community?" (Fig. 10.8) and "How do you communicate with neighbors? (multiple answers allowed)" (Fig. 10.9). Seventy-six percent of them answered that they have community meetings and 86 % of them answered that they have group work. These percentages are high, because the activities are necessary to sustain the community. Their main group activities are cutting grass, cleaning streets and shrines. Meetings and group works are community level activity and the result shows high rate in communication. Communicating with neighbors is individual level activity, and they have more choice whether they communicate with neighbors or not. Only 8 % answered that they do not communicate with neighbors often. As the result shows, they have close and daily communication in individual level activities also.

10.4 Possibility of Improving Information System

Since the environment surrounding the information recipients is complicated in case of disasters, it is preferable to have various information transmission media as much as possible (Usuda et al. 2009). Accordingly, disaster information receiving system has been focused and developed. Newspaper, radio and community wireless network are the major media which have been developed as media of disaster information. Recently the spread of internet makes it possible especially for young people to receive disaster information. Internet had an impact on not only issuing information from public organization but giving information from individuals. People will give disaster information through internet. However information transmission system from individuals is not developed yet. If disaster information from individuals are arranged and convened right way, the information is effective for DRR. In Reihoku area it is considered that the spread of internet especially among young generation is obvious. The internet environment is completely different since Sameura torrential rain disaster in 2004. Broadband was introduced last year, 2010, in Okawa village. An optical fiber cable is decided to be introduced in Motoyama and Tosa towns in 2010. Accordingly internet infrastructure is developing and the number of internet user will be increased in Reihoku area. Here information systems from public and individuals are summarized.

10.4.1 Necessity of Promoting Existing Information System from Public Organization

The questionnaire survey made it clear that TV and local newspaper are the major resources of both daily and disaster information in Reihoku area. Community wireless system is considered effective especially in case of disasters. Community wireless system is a risk communication tool. The system has been installed in each household for sharing disaster information originally. It is also used for sharing daily information as the questionnaire survey result shows. Residents obtain local information such as event information, weather and meetings, and it is a part of their life. If it is used only in case of disasters, they may not remember that they have it, they may forget where they put it and the instrument may be out of order. Since the system is used daily, it is maintained and used as same as in case of disasters. Since the community wireless system has been used for a long time and became old in Tosa town and Motoyama town, new system unifying an optical fiber cable is introduced in 2010 (Tosa Town 2010). Though community wireless system is useful when it become blackout, it is not sure that new system is useful when it become blackout. Therefore, having various communication tools is needed to secure and improve information system for disaster risk reduction. The questionnaire survey also showed the data of internet use. Though the rate of the use as disaster information resource is not high, the rate is high in younger generation. Therefore, more use in case of disasters is considered to be increased in the future.

In summary, younger generation is able to access more various media for getting disaster information because of the use of mobile phone, PC and internet in Reihoku area. Now internet and mobile phones play an important role as communication tools. Though the major purpose of possessing them is not for getting disaster information, it is used in case of disasters. Usually they are used for getting and giving daily information. Therefore they are a part of life for some users and more and more people have and use them daily. Since they are used daily, it is also used in emergency cases. There is a big difference between community wireless system, and internet and mobile phones. People are only able to get information from community wireless system, on the other hand, people are able to get and give information through internet and mobile phone. Most of younger generation have mobile phones and the possessing rate is almost 100 % as same as community wireless system. There are several disaster information services such as receiving disaster information automatically to email during disasters at charge and at no charge. Use of the service is one of effective ways for disaster risk reduction. It is simple and not difficult as an individual disaster risk reduction.

10.4.2 Necessity of Improving Information System from Individuals

Figure 10.10 shows the present transmitting information system. The vertical axis is for the publicness and horizontal axis is for quickness of delivering information after getting information. Before the distribution of mobile phone and internet, disaster information had been transmitted from public to private mainly, and the information system from public to private has been emphasized and developed. It is general that information collection and transmission is managed by disaster headquarters which government establish. In the information collection and transmission system, there is a process of information "arrangement" and "issuance" between the processes of information "collection" and "transmission". "Arrangement" means the evaluation of the authenticity and reliability of collected information. The "arranged" information is "issued" and "transmitted" by media (Usuda et al. 2009). Therefore, it takes for a while that collected information is transmitted. Though necessities of transmitting disaster information are accuracy, speed and appropriateness (Funatsu 1986), there is no case that all disaster information is collected and transmitted accurately, and lack of information and transmission is still a big issue (Usuda et al. 2009).

"The civil protection warning system" (J-ALERT) which notifies large-scale disasters, armed attack information and so on was constructed and the operation has started in 2007. J-ALERT delivers emergency information to the people via satellite; such as a huge earthquake will be observed or the announcement of evacuation will be issued. This system is expected to reduce the disaster victims and to prevent the

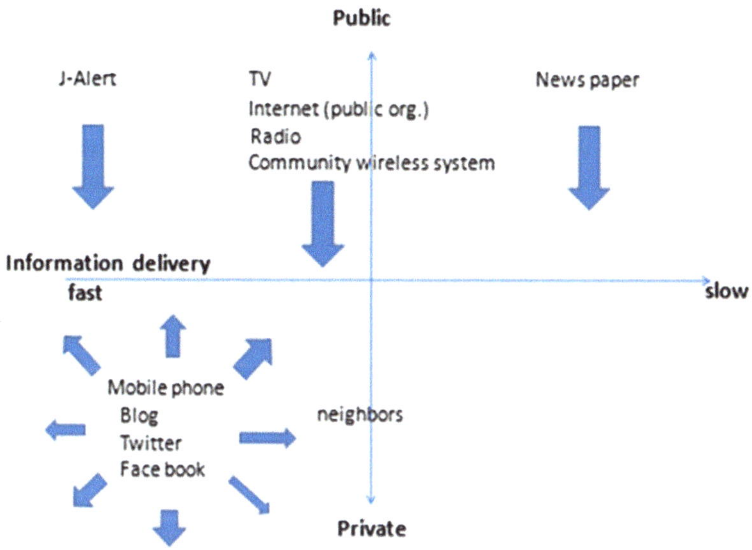

Fig. 10.10 Information system from public organizations and individuals

further disasters (Ohuchi and Izumi 2009). Since the information from the system is sent to people through contracted local community wireless system (Ohara et al. 2009), local offices in Reihoku area also are planning to make a contract.

After the distribution of mobile phone and internet, it is expected to develop variety of information system. Since the period of the use of mobile phone and internet is shorter than other media, the possessing rate becomes higher and higher especially for young generations. Since mobile phonc and internet are able to not only receive information but also provide information, they are expected to be used as information provider in case of disasters. Though the information from individuals is expected to be used in case of disasters, there are some characteristics to be considered. When mobile phone is used for receiving and providing information, the sender and the receiver know each other. However, information from individuals is provided through blog and social network system like Facebook and twitter, recipients of the information are not decided in many cases. Anyone is able to access the information, and recipients of the information are not decided. In addition, the number of information sender is numerous in case of blog and social network system. Since the media from public is limited, people have only a few choices of getting information and major information is shared. On the contrary, since the number of individual information sender is numerous in case of blogs and social network system, people have too much choice to access the information. Castillo et al. described users online, lack the clues that they have in the real world to assess the credibility of the information to which they are exposed. This is even more evident in the case of inexperienced users, which can be easily misleaded by unreliable information (Castillo et al. 2011). Therefore it is difficult to access necessary information.

Moreover there is no responsibility for individual senders if the information is correct or wrong. However, human communication and human network, which is always the fundamental for disaster risk reduction, is well developed, and people trust each other in Reihoku area, local disaster information provided by Reihoku residents is considered to be credible. It is needed the process of arrangement and making priority of importance in the transmitting information system from individuals. If there is a system and place to categorize credible information, people will access there. In case of Reihoku area, disaster information will be sent to reliable organizations such as fire volunteer and local governments, town offices and village office. Therefore, people who arrange information in fire volunteers and local governments are needed in case of disasters. Since the number of fire volunteer may not enough for arranging information, external help may be needed. The suitable people who arrange information in case of disasters are experts who know Reihoku area and disaster risk reduction. Since many university students visit Reihoku area as internship, their labs are considered to be one of the suitable external helps.

Acknowledgements The authors wish to thank Reihoku NPO for its support. The local information through them is always useful and questionnaires for the survey were thoroughly reviewed by the members. Without their support the survey was not implemented. In addition, this study would not have been possible without funding support by Sumitomoseimei.

References

Castillo C, Mendoza M, Poblete B (2011) Information credibility on twitter. In: International World Wide Web (WWW 2011) Conference Committee, Hyderabad, India. ACM, pp 675–684. ISBN 978-1-4503-0632-4/11/03

Funatsu M (1986) Disaster information and communication process. J Municipal Prob 38(6): 70–83 (in Japanese)

Gao H, Barbier G, Goolsby R (2011) Harnessing the crowdsourcing power of social media for disaster relief. IEEE Intell Syst 26(3):10–14

Gouda S (2007) Sediment-related disaster risk reduction and establishing safe village. In: Introduction to economics. The Open University of Japan (in Japanese). ISBN 4-595-23724-3

Ideta Ai (2012) Role of FM radio in post-tsunami recovery process in Natori, Miyagi Prefecture. Master Thesis, Kyoto University

Isono M (2009) Transmission and use of disaster information by ground digital data broadcasting. Japan Inform-Cult Soc 76–82 (in Japanese with English abstract)

Kanayama T (2007) Community ties and revitalization: the role of community radio in Japan. Keio Commun Rev (29). Available at http://www.mediacom.keio.ac.jp/publication/pdf2007/pdf/Tomoko%20KANAYAMA.pdf

Kochi Newspaper (2004) Kochi 2004, year end review. Kochi Newspaper (19 December 19, in Japanese)

Kochi Prefecture (2006) Comprehensive river basin management in Kochi prefecture. Kochi Prefecture (in Japanese)

Kochi Prefecture (2010) Outline of Reihoku forestry (in Japanese). 24 May 2010, from http://www.pref.kochi.lg.jp/~reihokurin/reihokusinrin/tukuru/2_12/2_12.htm

Kochi Prefecture and MLIT (2004) Case study of effectively functioned Sabo facility: 7. Kochi Prefecture and Shikoku Regional Development Bureau, Shikokusanchi Sabo Office, MLIT (in Japanese)

Meteorological Agency (2007) Climate Change Report 2008. Japan Meteorological Agency (in Japanese). Cited; available from http://www.data.kishou.go.jp/climate/riskmap/index.html

MIC (Ministry of Internal Affairs and Communications) (1985) Teletopia plan communication white paper. Available at http://www.soumu.go.jp/johotsusintokei/whitepaper/ja/s60/html/s60a01030100.html

MIC (1987) Council on the broadcasting in the age of new media (Broadcasting Policy Council) report

Nishimura T (2005) Experience of "Sameira torrential rain", Kochi prefecture. Chisui and Sabo 168 (in Japanese)

Ohuchi C, Izumi Y (2009) J-ALERT: the warning system which tells an emergency immediately. Res Rep Inst Electron Inform Commun Eng 109(117):141–146 (in Japanese with English abstract)

Ohara M, Jibiki Y, Sekiya N, Sum T, Meguro K, Tanaka A (2009) A survey report on effectiveness of earthquake early warning by J-ALERT system. Seisankenkyu 61(6):1055–1060 (in Japanese)

Otahara N (1991) Countermeasure against landslide in the Yoshino-river basin. Soil Mech Found Eng (Jpn Geotech Soc) 39(9):94–98 (in Japanese)

Peary B (2012) The role of social media in the 2011 East Japan earthquake and tsunami and its potential future application. Master Thesis, Kyoto University

Reihoku Administrative Office (2009) The second plan for Reihoku area (from 2005 to 2009) (in Japanese)

Shikoku Regional Development Bureau (2010) Retrieved May, 2010, from http://www.skr.mlit.go.jp/sabo/faciliti/faciliti.html

Tosa Town (2010) Development of information system in Motoyama and Tosa towns. Tosa Town Newsletter No. 443, pp 18–19 (in Japanese)

Usuda Y, Nagasaka T, Park WH (2009) A proposal of disaster information gathering and delivery system supported by regional collaboration and multi-media cooperation. Disaster Information No. 7 (Japan Society for Disaster Information Studies), pp 63–74 (in Japanese with English abstract)

Chapter 11
Community Resilience After Chuetsu Earthquake in 2004: Extinction or Relocation?

Miwa Abe and Rajib Shaw

Abstract The present study shows the findings obtained from the hearing survey conducted regarding relocation impacts and land use of relocated sites by the affected people of Chuetsu Earthquake, in 2004. The research targeted one of the affected communities named "Kotaka" which is located in Niigata Prefecture. As a result of relocation, people had a conflict within the community as well as with the external community. There also have issues within their own family members because of quick decision-making, community leaders, rumors and misunderstandings of the governmental relocation projects, as well as different opinions between young and old. However, through events such as sports festival, replacement of community temple, traditional dance and so on, it helped bonding better within the community. Kotaka's case study will show important factors that help to keep community identity after relocation. This way, relocation would not simply be just movement of the people (i.e. extinction) but movement of an entire community.

Keywords Community • Disaster recovery • Earthquake • Relocation • Settlement

11.1 Introduction

When the Hanshin Awaji Earthquake hit on January 17th, 1995, various measures were taken to support the victim's recovery such as daily life supporters and watchers who provided assistance especially to the elderly in the temporary or reconstruction

M. Abe (✉)
Kansai University, Osaka, Japan
e-mail: r118029@kansai-u.ac.jp

R. Shaw
Graduate School of Global Environmental Studies, Kyoto University, Kyoto, Japan
e-mail: shaw.rajib.5u@kyoto-u.ac.jp

R. Shaw (ed.), *Community Practices for Disaster Risk Reduction in Japan*, Disaster Risk Reduction, DOI 10.1007/978-4-431-54246-9_11, © Springer Japan 2014

accommodation. In contrast, in the case of the Chuetsu Earthquake of October 2004, because the affected area was in the depopulating mountain region; what was sought after was not mere recovery but long-term rebuilding of the area where the future is also taken into consideration. In order to provide human support for such efforts, the "community rebuilding supporter" was instituted in September 2007. The community rebuilding supporters are engaged with wide ranges of projects such as support for establishing networks for community rebuilding, co-ordination between the local residents and the local authorities, visiting victims to offer consultation as part of monitoring their welfare and providing information according to the problems of each area or the stages of rebuilding. In addition to maintaining human contact, the continuation of the "settlement" itself has been pointed out as an urgent issue. For instance, when some residents moved from the mountain region to the urban area because of the earthquake, this led to rebuilding from afar and some families were separated because the young generation with children moved to the urban area which is closer to schools and workplaces while the elderly stayed on in the familiar mountain region. Unlike the Hanshin Awaji Earthquake Disaster which hit the urban areas, the Chuetsu Earthquake has shown a new problem of rebuilding a community while dealing with depopulation in the mountain region. The current article focuses on collective relocation as one of the rebuilding projects. It tries to make sense of collective relocation which could lead to a rapid decline in population by examining settlements affected by the Chuetsu Earthquake. The chapter reviews the story of relocation and rebuilding as experienced by the residents of the settlements, a twin problem of depopulation and disaster.

It has been pointed out that an important aspect of post-earthquake rebuilding of a depopulating area is to draft a rebuilding plan that understands the relationship between different settlements and improves mutual help among settlements from a wider point of view (Kim and Yamazaki 2009). This suggests that it is necessary to comprehend the settlement structure from a wide and multi-layered point of view going beyond the affected settlements. Also, while depopulating settlements typically have problems with basic services necessary for daily life, a survey has found that around 90 % of the residents want to continue to live there (National and Regional Planning Bureau and Ministry of Land, Infrastructure, Transport and Tourism 2001). The survey further suggests that collective relocation should not be considered merely for post-disaster rebuilding purposes or because of depopulation. Research into post-disaster collective relocation has some issues such as how to make new human relationship, how to rebuild communities at the destination of relocation (Ikeda and Ishikawa 2006; Tanaka and Nakakita 2010), economic rebuilding and preparation of the site for relocation. However, all of these studies are focused on the "destination of relocation" and changes in the social environment such as the "site left by relocation", effects on the settlements around the site and post-relocation changes in the community activities are often overlooked. This research carried out interviews with residents who collectively relocated after the Chuetsu Earthquake of October 23rd, 2004 and the residents in nearby settlements before the relocation. It examined the changes in people's lives and the relationship with the original site where the settlement was originally situated. The article shows the ways in which the individuals and settlements made sense of the fact that they had to live in a new environment.

11.2 Disaster and Collective Relocation

11.2.1 Project for Improvement and Rebuilding

In relation to rebuilding efforts made after the Hanshin Awaji Earthquake Disaster and the Chuetsu Earthquake, a fundamental question i.e. "what is 'disaster recovery'?" was repeatedly asked. "Disaster recovery" was defined in the "Houmatsu Declaration" adopted at the launch memorial forum of the Association of Reconstruction Design in 2006: "disaster recovery means that people regains their spirit and vitality". Murosaki (2007) reviews various general and case-by-case discussions of reconstruction and shows the importance of sharing concrete aims. Inagaki (2007) emphasizes the importance of "staying close". While there was a trend to focus on processes in the past disaster recovery efforts, Nagamatsu (2008) suggests that while "staying close" may sound good, given that public funds are used to reconstruction, it is important to take moral responsibility towards the next generation. The current study focuses on specific issues of collective relocation. In Japan, the principle used to be "restoring the original state" but due to the catastrophic rain disaster of July 1972, an innovative system reform took place in which the principle of "improvement and restoration" was introduced: legislation on collective relocation (Act on Special Financial Support for Promoting Group Relocation for Disaster Mitigation).

This catastrophic rain disaster, which triggered the system reform, hit 32 prefectures from Tokhoku to Kyushu Regions with 433 people dead or missing (447 according to the Fire Department's White Paper), 6,855 buildings destroyed (of which 2,977 were totally destroyed and 10,204 were partially destroyed, according to the Fire Department's White Paper) and 5,987 landslides reported (Tomoda 1973). The Act was legislated in a very short period of time and although some problems were already clear right from the beginning, the intension was to prioritise reconstruction/rebuilding of the disaster victims' lives across the country. There were some settlements which carried out collective relocation before the Act was legislated including collective relocation of Iya settlement, Nagano Prefecture in 1961 (the number of household relocated: 235; the number of residents relocated: 1,323), and other instances in Nishitani village, Fukui Prefecture in 1965 and Kurokawa village, Niigata Prefecture in 1967, both of which were triggered by water disaster. Before the Act on Special Financial Support for Promoting Group Relocation for Disaster

Mitigation[1] was legislated, the collective relocation method adopted by Iya settlement, Nagano Prefecture was referred to as the "Iya method[2]" and emulated.

11.2.2 Group Relocation for Disaster Mitigation

The Act on Special Financial Support for Promoting Group Relocation for Disaster Mitigation was legislated at the 70th extraordinary Diet session of 1972 and it became possible to relocate residents before and after the disaster for counter-disaster purposes. In order to implement group relocation for disaster mitigation, the local authority is required to draft a group relocation for disaster mitigation plan and gain the approval from the Minister of Land, Infrastructure and Transport. It is also required to develop a residential area of a certain size with the lower limit of the number of relocating household[3] specified. The reason why the proportion of relocating household should be specified is to distinguish this type of relocation from the one that is carried out under "Special Financial Support for Relocation of Dangerous Buildings in the Near-Cliff Area"—which aims to help relocation of buildings from the disaster danger area. Under this scheme, the individual household can receive financial support but the subsidy rate is lower than in the case of group relocation for disaster mitigation. The plan has to cover buying agricultural land and residential land in the relocation promotion area as well as the land use plan including tree planting. Tomoda (1973) points out the importance of co-ordination among agricultural and forestry business in the nearby area which would be affected by the relocation. In implementing this project, attention should be paid to Article 4 on considerations by the local authority. The Article says "In drafting the group relocation for disaster mitigation plan, the local authority has to respect the wishes of the residents of the relocation promotion area and to pay attention that all residents in the relocation promotion area will be relocated." This means that not

[1] It aims to promote smooth collective relocation for disaster mitigation by providing subsidy for the project cost to the relevant local authority in order to promote collective relocation of housing in the area which is deemed to be unfit for habitation which is either an area where the disaster hit or is designated as a disaster danger area.

[2] When the disaster hit Iya, Nagano Prefecture, it was decided that residents' relocation to a safer place was preferable than carrying out recovery construction projects in the area. Instead of implementing support for recovery for the agricultural land, the national government's subsidy for the project was diverted from the then Ministry of Agriculture to the Ministry of Home affairs and paid out as subsidy for collective relocation.

[3] The residential area of a certain size refers to the residential development of a size which can accommodate more than 10 houses when fewer than 20 houses relocate, and more than half the number of house that need to relocate when more than 20 houses relocate (Act Article 2, Clause2, Order Article 1, Rule Article 1). In the case of the Chuetsu Earthquake of October 23rd, 2004, the required size of the residential development was lowered from ten to five houses in consideration of the depopulating settlements in the mountain areas. An additional measure such as applying a higher basic sum than in general was also taken (implemented on April 1st, 2005).

all households in the relocation promotion area have to relocate. Depending on the interpretation, the article allows some disaster victims to stay even after the disaster if the consensus on relocation is not achieved.

Usually, the disaster hit settlement relocates to a nearby flat land after developing it. However, the collective relocation of Himeko town,[4] Amakusa county of Kumamoto Prefecture, one of the settlements affected by the heavy rain of July 1972, was unusual. Himeko town's damage was 45 dead or missing, 74 injured and 162 buildings destroyed. Himeko town experienced flooding and landslides of various sizes and they developed a new residential area by reclaiming from the sea.[5] In March 1976, 178 households relocated under the group relocation for disaster mitigation project (Kinoshita 1984). Between 1972 and 2006, a total of 35 local authorities and 1,834 households relocated under the group relocation for disaster mitigation projects (Ministry of Land, Infrastructure and Transport 2012[6]).

11.2.3 Collective Relocation: Overseas Examples

In turning to overseas cases of big disasters in the past including the Flores earthquake, Indonesia (1992), Hanshin Awaji Earthquake Disaster (1995) and tsunami in Papua New Guinea (1998), many disaster victims were forced to relocate from their familiar environment. In the case of Papua New Guinea, all the coastal settlements were relocated to inland (Maki et al. 1999). However, as a part of post-disaster resettlement involving relocation, gathering several communities in one site based on different living environments and religions is generally avoided because of the difficulty of forming a new community and weak sustainability of such a community with households leaving the new settlement as time passes. Also it has been pointed out that when environmental relocation happens with the relocation of life base the residents are forced to reconstruct their relationship with the environment which has been developed over a long period (Iwaisako et al. 1996). While a variety of problems with collective relocation have been identified, when looking at the case of the Indian Ocean Tsunami of 2004, it turns out that many coastal areas hit by the Tsunami were eroded and not only the buildings but also the land itself was lost rendering human inhabitation impossible; because the government designated areas where residence was banned, the continuation of life in the original site was politically closed; and there was much aid for relocation from abroad. In Sri Lanka, the Japanese government was one of the supporters of relocation. Problems associated with changes in community culture due to collective relocation and sustainability of such culture have also been identified suggesting that the challenge lies not in the mere act of relocating buildings but in maintaining the community as a group.

---

[4] It is now part of Amakusa-shi, Kumamoto Prefecture as a result of merger on March 31st, 2001.

[5] Four sites were reclaimed from the sea. A total of 178 households from 28 sites relocated to the total relocation area of 62,000 m^2.

[6] http://www.mlit.go.jp/crd/chisei/boushuu/boushujoukyou.pdf (accessed on October 30th, 2012).

11.3 The Outline of Chuetsu Earthquake and Affected Settlement

Kawaguchi town was a part of Kita Uonuma County, Niigata Prefecture until it was merged with Nagaoka city on March 31st, 2010. Because it is not located adjacent to Nagaoka city, it consists a detached territory of the city. Kawaguchi town is located at the confluence of Shinano River and its branch, Uono River. The main industry is production of the Koshihikari brand of rice from Uonuma and varicolored carps. In 2004, 9 years after the Hanshin Awaji Earthquake, a major earthquake of magnitude of 6.8 hit Kawaguchi town.

At 17.56 on October 23rd, 2004, an earthquake with the epicentre in Chuetsu region of Niigata Prefecture measuring the magnitude of 6.8 occurred and brought about major damages including 48 dead, about 100,000 evacuees, 90,000 houses destroyed and the total damage reaching 3,000 billion yen (Niigata Prefecture 2005). In addition to damages to the buildings caused by the earthquake, many landslides that were triggered by it brought serious damages to both human and material aspects. The damage caused by the Chuetsu Earthquake is summarised in Table 11.1.

Kotaka had located in Tamugiyama area before relocation. Interviews were carried out with six people. Three interviewees are Kotaka settlement which implemented collective relocation. Two interviewees are Tamugiyama residents which was the original site of Kotaka. And one interviewee is a staff member of the Kawaguchi town hall at the time when collective relocation was decided.

When the earthquake hit, there were 104 residents in 25 households in Kotaka area of Kawaguchi town.[7] It was situated in Tamugiyama area of Kawaguchi town and it was located farthest in the mountain since Yamano-aikawa settlement, then the farthest community in the area, had relocated in 1971. The settlement's location before and after collective relocation is shown in Fig. 11.1.

Table 11.1 Damages caused by the Chuetsu Earthquake

Casualties (number of people)				Damages to houses (number of houses)			
Dead	Missing	Serious injury	Minor injury	Totally destroyed	Mostly destroyed	Half destroyed	Partially destroyed
48	0	634	4,160	3,173	2,144	11,559	103,495
Damages to non-residential buildings (number of buildings)		Other damages (number of sites)					
Public facilities and others		Roads		Rivers		Landslides and others	
40,368		6,064		229		442	

Source: Sediment Control, Construction Department, Niigata Prefecture, published on August 4th 2005

[7] Merged with Nagaoka-shi on March 31st, 2010.

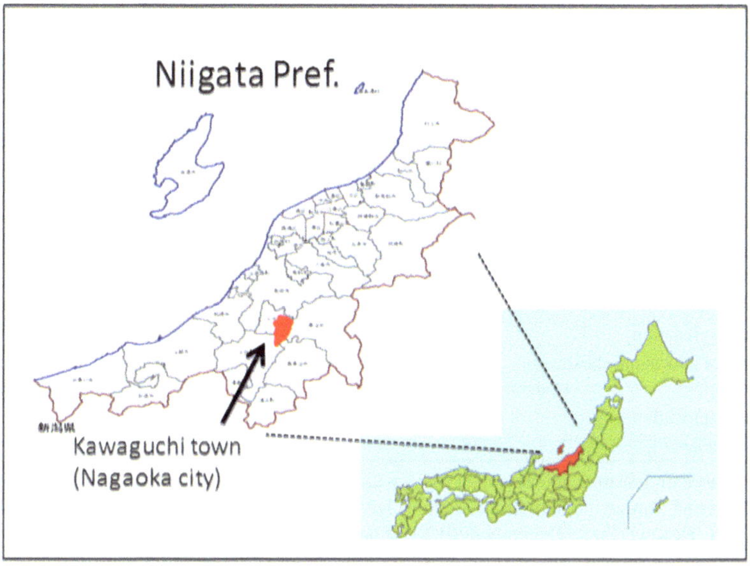

Fig. 11.1 The location of Kawaguchi town (now Nagaoka city)

11.4 Areas Affected by the Chuetsu Earthquake and Collective Relocation

11.4.1 Details of Collective Relocation

In terms of collective relocation with a view to revive the settlement caused by the Chuetsu Earthquake of 2004, several areas implemented collective relocation by combining various support measures from the national government so as to reflect as many residents's view as possible, especially the elderly residents who would find it very difficult to rebuild their lives on their own (Table 11.2). Kawaguchi town authority made the decision to draft promotion of group relocation for disaster mitigation for Kotaka settlement by May 25th, 2005 and decided to develop a residental area for 19 households in Iwadehara area of Nishikagaguchi of Kawaguchi town. Kotaka settlement was the place where the evacuation recommendation was lifted last in Kagaguchi town. A month after the earthquake, in November, residents had decided to relocate collectively and informed the local authority of their intension. Originally, 19 households wanted to relocate collectively and further six households were due to leave the settlement by moving out to a different area or joining the children's household elsewhere. In reality, as of December 2012, 18 households relocated collectively to Iwadehara and 1 household remained in Kotaka settlement.

Table 11.2 Implementation of projects for promotion of group relocation for disaster mitigation caused by the Chuetsu Earthquake

Implementation body	Nagaoka city		Ochigaya city		Kawaguchi town
Decision made	12.26.2005	12.26.2005	12.26.2005	09.21.2005	07.12.2005
Relocation promotion area (settlement)	Nishitani area (Etsuro)	Four areas (Rangi, Nigoro, Kubizawa, Asahi)	Shioya area	Junidaira area	Kotaka area
Implementation period	2005–2006 (2 years)				
Original number of households	159	129	49	11	25
Relocated household	16	45	24	11	24
Collective relocation	13	31	15	10	18
The name of relocation destination	Nishitani Danchi	Sabusho danchi, Chigaya Danchi		Sabusho Danchi	Iwadehara Danchi
Individual relocation	3	14	9	1	6
Ratio of leavers (%)	10	35	49	100	96

Source: Based on Ishizuka and Sawada (2010) with the author's amendments (according to the table in Ishizuka and Sawada (2010), 25 households left Kotaka settlement with 19 households relocated collectively. However, as of December 2012, one household has given up on relocation and therefore the number of households relocated is 24 with 18 households collectively relocated)

Collective relocation of Kotaka settlement was not a relocation to the nearby land, and they moved to an area 4 km away from the original site. In terms of project implementation, "special financial support for promoting group relocation for disaster mitigation" was used because of its favourable subsidiary rate and "support for improvement of small-scale residential areas" was used to develop public housing to provide for four households which could not obtain mortgage due to the householders' age. Because of the use of the support for promoting group relocation, the original settlement was then designated as the relocation promotion area and no one was allowed to go back to live there. In contrast to Kotaka settlement, former Yamakoshi village, chose the support for improvement of small-scale residential areas in order to preserve the possibility of rebuilding the original settlement (Ishikawa et al. 2007).

Junidaira area of Ochigatani city which was forced to relocate collectively by the same Chuetsu Earthquake is known as a successful example of collective relocation. According to Fukutome (2011), this was due to (a) all 11 households were members of the building refurbishment mutual aid scheme offered by the agricultural co-operative and everyone received the full payout; (b) the land price went down because of the subsidy provided by public projects related to reconstruction and (c) all destroyed houses were more than 25 years old and their mortgages had been paid off. However, in the case of Kotaka settlement, there was a household which rebuilt its house just before the 2004 earthquake, and it left the collective relocation scheme in order to avoid having double mortgage obligations. In the case of collective relocation of Kotaka settlement, there were a number of factors including a strong leader, the diverging opinion about relocation between generations who shared the same house.

11.4.2 Other Collective Relocations in Kawaguchi Town

There are two settlements in Kawaguchi town which implemented collective relocation. Yamano-aiwaka settlement which implemented collective relocation in 1973. Driven by the heavy snow which brought serious damages to Eastern Japan from 1963 to 1964[8] and the rapid depopulation, it collectively relocated to the present location, Nishi-kawaguchi area. The residents recalled "collective relocation had already been talked about in the settlement several times" and "because of the push in the form of authority's assistance, relocation took place in the end". The temporary law for 10 years, the "Act on Emergency Measures for Dealing with Depopulation" legislated in 1970 became the driving force. The local authority chose the relocation destination and developed a residential area and the town hall bought up the privately owned housing land and agricultural land and fund raised by this was used to pay for the relocation. There are still some households that retain the ownership of the land they lived on before the relocation, but most of the residents had their land bought up by the town hall excluding forests. Yamano-aikawa settlement was situated near the border[9] between Kawaguchi town and its adjacent municipality (Horikoshi) and some residents thought that the local authority implemented collective relocation in order to prevent a further decline in population as many residents had moved to Horikoshi town. It is clear that Yamano-aikawa settlement, located farthest in the mountains, saw a rapid decline in population in the first half of 1970s (Fig. 11.2). All settlements in Tamugiyama area had experienced a long-term population decline although the number of household remained relatively stable.

Kotaka settlement was situated further downstream than Yamano-aikawa settlement and after Yamano-aikawa settlement collectively relocated, it became the "dead end[10]" settlement. Kotaka settlement also decided to relocate collectively after the Chuetsu Earthquake of 2004 and moved to Nishi-kawaguchi where Yamano-aikawa settlement had relocated. As a result, after about 30 years since Yamano-aikawa settlement's relocation, the two settlement became neighbours again.

Kotaka settlement also had heavy snow every year but because it was not as high as Yamano-aikawa settlement, it had access to the urban area by car. In order to implement collective relocation with the support from the national government for projects to promote group relocation for disaster mitigation, it needs to be officially designated as hazardous area. Previous Kotaka settlement was designated to be the

[8] Often known as "the Three Eight Heavy Snow", and it is officially called "the Heavy Snow of January 1963 ('Sanpachi Gousetsu')". There was a heavy snow fall for about a month from the end of December 1962 to the beginning of February 1963 over a wide area from Tohoku to Kyushu region with the heaviest fell on Hokuriku Region. The Fire and Disaster Management Agency. Nagaoka city, Niigata Prefecture, recorded the maximum snow fall of 318 cm.

[9] The settlement was situated near the mountain top and it was easier to get to Horinouchi, the neighboring town over the mountain than to get to the city centre of Kawaguchi town.

[10] The term was used by the residents in the interviews.

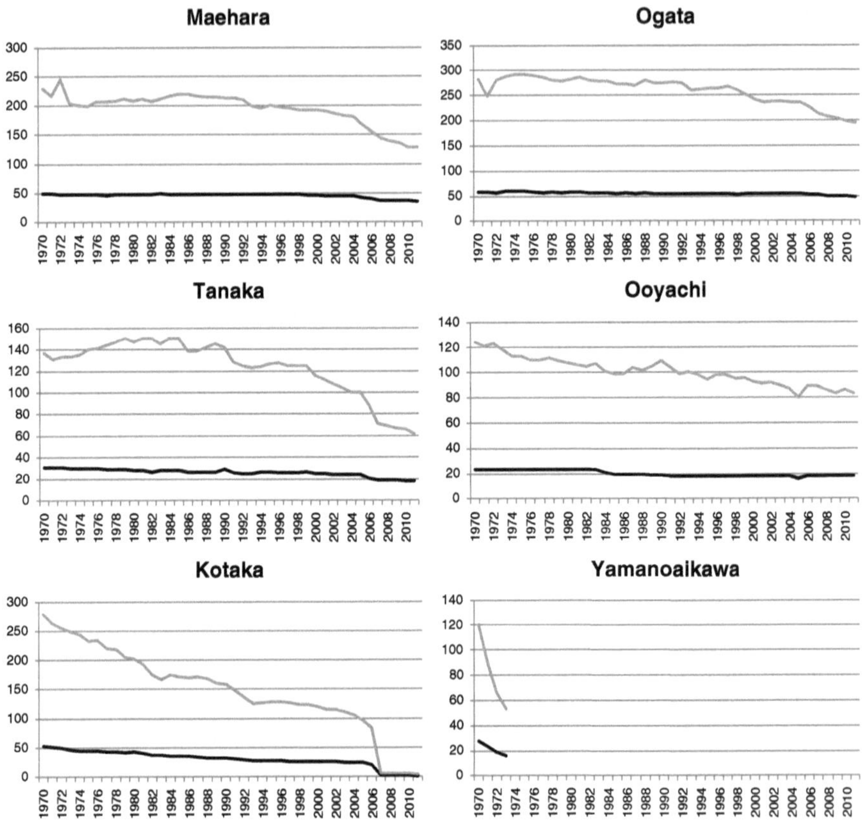

Fig. 11.2 Changes in the number of household and population in Tamugiyama area between 1970 and 2011 (*black line*: the number of household, *gray line*: population) (created by the author based on data from Kawaguchi Branch of Nagaoka city)

area where secondary damage by the Chuetsu Earthquake occurred. The mudflow had stopped the river from flowing by forming a dam and when it bursts the settlement would be exposed to the danger of debris flow. Instead the former home remained designated as a danger area and the residents cannot go back and live there, people can get subsidy.

11.5 Tensions Caused by the Relocation (Findings from the Interviews)

11.5.1 Residents' Emotion

First interview was carried out in Kotaka new settlement on November 11th, 2009. It collected various views such as "collective relocation had been discussed at an early stage" and "partly we were tempted by money". There was an episode with a particularly strong impression. While the first author was doing interview with Ms A, she told that the next interview will be with Ms B. Hearing that, Ms A suggested to invite Ms B to her house for the interview. Ms A explained that since the relocation, communication with Ms B had been sparce. This message was sent to Ms B without reflecting on its implications. When Ms B heard about this, she asked repeatedly "did she really say so?" She then prepared a little present and headed to Ms A's house. It was a very cold day and the interview with Ms A was carried out with both of us in the kotatsu (the Japanese foot warmer). Interview was conducted around the kotatsu at the same position as before. But Ms B sat on her heels on the cold tatami mats and would not put her feet in the kotatsu. Ms A was preparing tea in the kitchen. Ms B was waiting for Ms A to return without putting her feet in the kotatsu. As soon as Ms A, who came back with a pot of tea, said "don't be silly, do get in the kotatsu", Ms B bowed deeply while still sitting on her heels on the tatami mat saying "I am really sorry that we caused a lot of trouble at the time of relocation". It turned out that Ms B's son was one of settlement leader when the decision to relocate collectively was taken. And Ms B felt guilty that the relationship among the settlement residents became tensed due to the relocation. Ms A responded "a lot has happened but they are all in the past", "let's be just like before" and Ms B finally put her feet in the kotatsu. This episode was a very short exchange but for Ms A and Ms B. But it was the first time they could talk face to face 4 years after the earthquake. It is one of relocation impact to Kotaka.

The effect does not only affect on the relationship among the residents. In the previous site, the houses were spread out in the settlement and there was some distance between the houses. However, the residential area after relocation is designed to be housing complex and the distance between houses is short. Therefore if people open the window, they can see the neighbour's window. So people do not open the window often. For example, when they hear someone is taking a shower, they feel embarrassed and close the window. These were the examples cited by the residents that they were more conscious of everyday noise from next door which they had not been bother at all in the pre-relocation site. A sense of embarrassment which makes them very constraint of the next door neighbours even in the house. This was not experienced before the relocation.

11.5.2 The Relationship with the Land in the Pre-relocation Site

Mr C, who was the leader when the residents started to discuss collective relocation at the evacuation center, negotiated an early collective relocation on behalf of the residents. Some elderly residents wanted to return to the familiar place rather than to relocate, but some others did not want to return to the place under heavy snow, and some wanted to relocate so that children's schools would be closer. The residents were indecisive between relocation or restoring the original site. In the end, the decision was reached that Kotaka settlement should be relocated as a settlement. After that, five representatives were engaged with negotiation with the local authority and the residents in order to implement relocation and they also looked for the place to relocate. However, as time went by, the view changed to the one that relocation was a rushed decision. Above all, there were a few people whose houses were still standing in Kotaka settlement. Surely the earthquake messed up inside the house but the houses themselves were still standing. There were two elderly ladies who commuted back to their houses by community bus or on foot every day after the relocation. They went back not to tend the field but "to feed the cats". The residents of Tamugiyama area felt "sad" and "sorry" whenever they saw these elderly ladies. Others commuted back to tend their fields. The former houses were used as "sheds". When they saw the house itself was still standing, they seem to think we want to go back. The same goes for the grave. With the collective relocation, some families also relocated their ancestral graves. There were, however, residents who gave up on relocating their ancestral graves because they had no heirs or they did not have enough funds to cover grave relocation. When the first *higan* (the equinoctial week) came after the relocation, for the first time they could think about the grave: "well, the house has been relocated but what about the grave?" Until then, they were preoccupied with rebuilding their lives. When the life became a little calmer, they realised it was not enough to relocate the house and started to wonder: "what are we going to do about our ancestors?"

The issue of construction of an industrial waste disposal factory came up in relation to the original site. Some Kodaka residents tried to sell their land knowing that they were bought to build an industrial waste disposal factory so as to pay for house building or buying land to rebuild their lives. This was a decision made to support their lives and it was an option in order to live for the Kotaka residents. On the other hand, as in many places in Niigata which is known for its rice production, many residents in Kawaguchi town earned their living by rice cultivation. The residents of Tamugiyama area opposed to the idea of building an industrial waste disposal factory in the upper stream area because it would affect their livelyhood in the lower down the stream. Those residents of the lower stream rejected the construction of an industrial waste disposal factory and started to hold group meetings and carried out inspection tours to the industrial waste disposal factory in other areas in order to continue their opposition. In the end, the plan was ditched but some residents in Tanamugiyama area where in the original Kotaka settlement was situated and that they were unhappy: "we thought they were from the same village", "they do not worry about the place they have left". As a result, tension between those who sent off Kotaka residents and those who relocated emerged. Figure 11.3 shows some glimpses of before and after the disaster.

11.5.3 The Symbol of "Kotaka": Sports Day and Yakushido

Kotaka settlement used to hold a Sports Day on September 23rd every year. It was held regardless of the weather and it was a major event in the community in which everyone from children to adults took part. The Sports Day was suspended for

Fig. 11.3 Housing condition before/after relocation. (**a**) Before relocation site, (**b**) after relocation (housing complex), (**c**) message from relocated people. At the flag (5 years passed from earthquake. We live life to the extreme), (**d**) message for cerebration of completed "Yakushi-do"

1 year after the earthquake, but sports day for residents was resumed 1 year after the relocation. Up to that time, people were concern that relationships with neighbors were broken due to the relocation. In fact, brawls happened at the residents' meetings and some residents did not greet each other, all of these contributed to the mounting concern over rebuilding neighborly relationships among the residents. However, in the 1st year of relocation, the traditional residents' Sports Day was resumed. September 23rd happened to fall in the middle of the rice harvest season. Although this is the major event, the residents would feel "if it is fine, I'd rather harvest rice" or "I am too busy to join the Sports Day". They joined the event when "it is raining and I cannot do harvesting, so this is a good timing". They used to have celebrations in the settlement twice a year: the spring celebration and the fall celebration (Yakushi celebration). However, because of the decline in population and because of overlap with the busiest time in agriculture, only one celebration a year was now taking place and the date has also changed. However, as far as the Sports Day is concerned, the change of date was repeatedly discussed to avoid overlap with the rice harvest season. At the time, the leader would refuse to change the date by saying "it cannot happen during my term" and the event continues and the date remains the same up to today. The origin of the event appears to be the teachers' efforts to involve local residents due to the decline in the number of elementary school students.

Its history is therefore much shorter than that of the Yakushi celebration. But for the residents, the Sports Day is more important than the other local festivals. It is an all day commitment, the residents say. They start preparation for the day on the previous day to enjoy competition until afternoon. This would be followed by a post-evening partly and the residents would spend time together all day eating and drinking. Compared to the original site, the current relocated site is much smaller. The residents, however, are using roads around their houses and community spaces for the event. People say that "even our site was smaller than before, but we can enjoy with small site". No other settlement in Kawaguchi town has its own Sports Day. The Kotaka residents say that if we lose our community's Sports Day it would no longer be Kotaka".

Also, the Yakushido (the shrine to Yakushi Buddha) was relocated with the settlement. More precisely, because the Yakushido itself was damaged by the earthquake, the statue in the shrine was relocated to the new site. Just after the earthquake, the statue was with the residents in the evacuation centre, and when they moved to temporary accommodation, the statue was stored in the storage space in the original site which was not much damaged by the earthquake. When almost all residents relocated, a wish to take the statue with them was expressed. The Yakushido which used to stand on a little hill in the settlement and had a commanding view of the houses and fields, was destroyed by the earthquake. There were ancestral graves of the Kotaka resident around the Yakushido. Some of the headstones fell down and some of the graves were removed by the landslides. The Yakushido used to be the place where the residents would drop by when they came to the grave and when they visited the Yakushido, they would stop by at their ancestral graves. In the original site, the Yakushido served as a "place" where the residents got together. They came to the Yakushido for their first visit in the year in January. The Yakushido and its front yard were used for the settlement's events such as a once-a-month tea party for women and celebrations. The Yakushido offered an additional place to get together for the local residents apart from the school and community center.

As rebuilding gathered pace, relocation of the Yakushido was also discussed. It turned out that rebuilding fund could be used to build a new Yakushido, but it could not be used to buy the land on which to build a new Yakushido. It was then decided to use the savings of the local residents' association to buy the necessary land.

11.5.4 Resumption of Celebrations/Kotaka Stick Dance

The former Yakushido was a big building and had enough space for people to sit down having a chat and tea. The new Yakushido is small and has enough space for the statue only. However, the new Yakushido has managed to secure a front yard which is used to practice and demonstrate Kotaka Stick Dance. Kotaka Stick Dance was not performed for a while but the Kotaka Stick Dance Preservation Society has been set up and the art is now handed down to the next generation. The particularly interesting thing is the society's membership. The head of the only household that

Fig. 11.4 Yakushido and Kotaka Stick Dance. (**a**) Yakushido after Chuetsu Earthquake, (**b**) new Yakushido and stick dance

gave up on relocation and has stayed on the original land is a member of the society. Given that some complaint against the leader and regret about the subsidy for relocation has been expressed by the residents, it is natural to assume that the process of forming a consensus on relocation was not necessarily smooth and trouble-free. However, contact between the remaining household and the relocated households have been maintained through Kotaka Stick Dance, the community's traditional art, even after the relocation. The knowledge of the danced style of Kotaka Stick Dance has been preserved but the knowledge of accompanying music was long lost and no resident knows what music to play. At present, efforts are made to revive the music by asking for help from other settlements which have similar music with assistance from the community rebuilding supporters (Fig. 11.4).

Six dancers are involved in Kotaka Stick Dance. If one dancer is missing, there is no Kotaka Stick Dance. The initial concern was whether they could hand down the accompanying music to the next generation even if it has been revived. A session to teach the settlement's children Kotaka Stick Dance was scheduled on the day before the earthquake hit. After the earthquake, no one knew to what extend the dance could be revived after relocation. In 2011, two young people from the area joined the society. On October 23rd, 2011, when the "Chuetsu Memorial Corridor: Kawaguchi Kizuna Kan" opened, all settlements in Kawaguchi town took part in the relay marathon "Kizuna Marathon" and demonstrated their own traditional art. The Kotaka Stick Dance Preservation Society with the two new members demonstrated their dance to a large audience.

11.6 Conclusion

Kotaka settlement was designated as the relocation promotion area and almost all households relocated from the mountain area to the flat land. However, fields and uninhabited houses are left in the original site. Some houses which were partially destroyed by the earthquake had garages which were used for evacuation after the earthquake and they can be used for an overnight stay now. Some residents would say "when I look at the house…", "I'd like to go back" when they see the place where their lands and houses remain but life can no longer be rebuilt.

A similar case was reported in connection with the relocation project after the water disaster of July 1972. Kawabe town, Akita Prefecture (Fukuda area, Kawabe village, Kawabe County, Akita Prefecture) was a remote settlement with 59 people in 11 households, with houses built on the both shores of Kan-nai River. The river flooded due to the heavy rain on July 31st and all 11 households relocated 6 km downstream after the disaster. One-third of the agricultural land in Fukuda area was damaged but it was rebuilt and the residents kept tendering the fields by commuting there without asking the local authority to buy up the land. The houses were also left alone without being demolished and it was left for the nature to take its course (Nishizaki 1973). When the house is left behind, however far they have relocated, they would feel "I'd like to go back". This is because there is not only the "house" but also memories of the time which they spent with their family in the house. It may well be necessary for people to relocate to a safe place (or a place where less future damage is predicted) in the process of post-disaster rebuilding. However, the decision to relocate, especially in Japan, to relocate collectively using the project to promote group relocation for disaster mitigation means that they have to accept the condition that they could never return to the place they lived up to then in exchange for a life in a safer place. Considering them does recovery/rebuilding has to be "fast"?

Among works on the bereavement of the residents who were forced to relocate, Fried (1963) regards attachment to the place as "the tie that is recognized when it is lost" and suggests that the theory of attachment in interpersonal relationship can be applied. Perhaps, the function of the "safe haven" in interpersonal relationship suggested in the attachment theory by Bowlby (1969) and Ainsworth et al. (1978) could be replaced by the "former land" in the words of the disaster victims. However, those who have relocated have to continue living. We need to consider what value they find in the new place and what support can be provided to facilitate this. The cost for relocation is surely the major issue. But what we learn from the examples in Kotaka is that while the former site could be a "former place" for those who have relocated, for those people who are staying there it is a resource that has close relationship with their life, that it takes time to rebuilt relationship among the residents who have relocated and that there are challenges which would emerge when the life becomes calmer such as the ancestral graves and communal celebrations (Yakushi celebration). What we can do, then, is to share these examples of problems the disaster victims may face as they continue to rebuild their lives and to think what could be done together.

In this study, three new relocation issues were showed (Fig. 11.5). First one has relationships with previous neighboring community and land. Usually, previous relocation studies have showed relocated site's issues, but this result has points the necessary of considering with previous area. Second is a sense of embarrassment which was not experienced before the relocation such as sound from next houses.

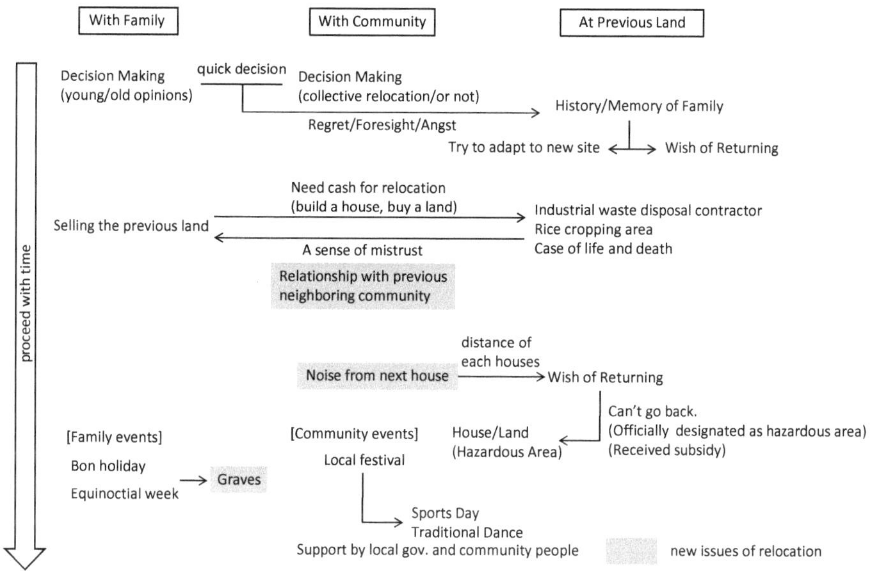

Fig. 11.5 Relocation issues with time in Kotaka

Third is graves issue. For Japanese, grave is an important thing. However just after relocation, people did forget their graves. When people remembered about it, nobody received any support and they understood own situation.

The Kotaka residents are trying to find value in the new site in the midst of policy-led rebuilding project called collective relocation while still having a wish to "go back to the former land". On the other hand, by maintaining and reconsidering the ties and relationship with people in the place they lived for a long time, they are trying to "maintain" the village in a new place and not engaged with the "demolition" of the settlement by the earthquake and relocation.

Acknowledgement This is a product of research which was financially supported in part by the Kansai University Subsidy for Supporting Young Scholars, 2012. "Social Vulnerability Analysis of Households in Kita-ku, Osaka". And I would like to express the deepest appreciation to people in Kotaka, Niigata Prefecture.

References

Ainsworth MDS, Blehar MC, Wasters F, Wall S (1978) Patterns of attachment. A psychological study of the strange situation. Lawrence Erlbaum Associates, Hillsdale

Bowlby J (1969) Attachment and loss, vol 1. Plenum, New York (attachment)

Fried M (1963) Grieving for lost home. In: Duhl LJ (ed) The urban condition: people and policy in the metropolis. Simon & Schuster, New York

Fukutome K (2011) This paper was reported at Disaster Area Residents Exchange Organization in Jan 16–17, 2011. People can direct access to this paper from [http://www.pdhsk.com/hukutome.pdf] or HP (of Disaster Area Residents Exchange Organization) from [http://www.pdhsk.com/shusi.htm]

Ikeda H, Ishikawa E (2006) Characteristics and issues of group removal project for disaster prevention as a system for reconstruction of settlement in depopulated mountainous area. Summary of technical papers of Institute of Social Safety Science 18:28–24

Inagaki F (2007) Hisaisha ni yorisou chukan shien soshiki wo [Establishment of intermediate support for affected people]. Fukkou Dezain Kenkyu [Disaster recovery design study], Special Edition, p 3

Ishikawa E, Nakagawa I, Ikeda H (2007) Recovery process of housing and suffers' lives in landslide areas: recovery project with group relocation and landslide predictions after the Niigata-ken Chuetsu Earthquake. Collection Papers of Housing Research Foundation 34:303–314

Ishizuka N, Sawada M (2010) A study on restructuring a community with promoted group relocation for disaster mitigation – a case study of Jyunidaira, the relocated town of Ojiya City hit by Niigata Chuetsu Earthquake. Summary Tech Pap Inst Soc Saf Sci 26:1–2

Iwaisako H, Miura K, Maki N, Kobayashi M (1996) Relocation to restoration housing due to the South-West off Hokkaido Earthquake: Natural Disaster and Environment Transition No.2, 137–138

Kim D, Yamazaki J (2009) Depopulation and village characteristic of Noto Peninsula earthquake stricken area region: attention to population and number of households change in 1960 and 2005 of which the unit is the old village. Collection Papers of Architectural Institute of Japan Kinki Branch. Planning 49:233–236

Kinoshita K (1984) Shouwa 47nen sai kamiamakusa daisaigai to shudaniten no sonogo (Collective relocation and the after with a major disaster at Kamiamakusa in 1972). Kikan shoubou (Q Issue Disaster Prev) 74:108–118

Maki N, Hayashi I, Hayashi H (1999) Disaster management in the Papua New Guinea tsunami disaster on 17 July, 1998; Recovery process of the social flows and stocks from disaster. J Soc Saf Sci 1:195–200

Ministry of Land, Infrastructure and Transport (2012), http://www.mlit.go.jp/crd/chisei/boushuu/boushujoukyou.pdf. Accessed 30 Oct 2012

Murosaki Y (2007) Murosaki sensei ni Fukkou ni tsuite kiku [Question of disaster recovery to Prof. Murosaki]. Fukkou Dezain Kenkyu [Disaster Recovery Design Study], special edition. pp 1–3

Nagamatsu S (2008) Fukkouron no heisokukan wo dou norikoeru ka (How to overcome form difficulties of disaster recovery). Fukkou Dezain Kenkyu (Disaster Recovery Des Study) 8:2 [Fukkou ron no heisokukan wo dou norikoeruka (How to overcome a sense of stagnation from disaster recovery)]

National and Regional Planning Bureau, Ministry of Land, Infrastructure, Transport and Tourism (2001) Kasoshuraku kenkyukai houkokusho (gaiyou) [The report of depopulation community (Summary)]

Nishizaki M (1973) 47nen hassei daikibo saigai to shuudan iten (Wide-scale disaster and collective relocation in 1972). Kikan shoubou (Q Issue Disaster Prev) 44:19–28

Sediment Control, Civil Engineering Division, Niigata Prefecture (2005) Niigata ken chuetsu jishin to dosha saigai [The mid Niigata Prefecture earthquake of 2004 and sediment disaster]. http://www.pref.niigata.lg.jp/HTML_Article/864/780/chuetsujisin.pdf#search=%27%E4%B8%AD%E8%B6%8A%E5%9C%B0%E9%9C%87+%E5%B0%8F%E9%AB%98+%E5%9C%B0%E5%9B%B3%27. Accessed 17 Dec 2012

Tanaka M, Nakakita E (2010) The realities of separation of disaster-stricken villages by the collective relocation project ant its impact on residents: case of Nishidani District, Nagaoka City, Niigata Prefecture. Collection Pap Inst Soc Saf Sci 12(13):463–470

The Fire and Disaster Management Agency, Ministry of Internal Affairs and Communications. Shoubou Nenpyou [Chronology of the fire and Disaster Management Agency]. http://www.fdma.go.jp/html/hakusho/h19/history/index.html. Accessed 22 Aug 2012

Tomoda N (1973) Bousai no tameno shudanitensokushinjigyou ni kakaru kuni no zaiseijyou no tokubetsusochitou ni kansuru houritu ni tsuite (The low of special financial support for promoting group relocation for disaster management). Kikan shoubou (Q Issue Disaster Prev) 43:5–10

Chapter 12
Issues and Challenges in Temporary Housing in Post-3.11 Kesennuma

Yasutaka Ueda and Rajib Shaw

Abstract In the aftermath of the Great East Japan Earthquake and Tsunami (EJET), which caused enormous damages, more than 53,000 temporary housing units were provided by the government based on the law. It is mentioned in the first half of this chapter that care enhancement activities and promoting the development of a residents' association are vital in temporary housing due to features of residents and habitation in temporary housing with low quality from the results of surveys conducted by Ministry of Health, Labour and Welfare. The last half of this chapter analyzes key lessons of community development in temporary housing in the case of Kesennuma City. The results indicate that creating communication opportunities to share problems and establish relationships with the municipality and specialists for solving the problems is crucial to community development. In addition, needs in the fields of medical care, health care, welfare, and intergenerational exchange increased in temporary housing sites.

Keywords Community development • Residents' association • Temporary housing • The Great East Japan Earthquake and Tsunami (EJET)

12.1 Introduction

In Japan, temporary housing is provided by the government based on the law after a large-scale disaster happens. Accordingly more than 53,000 temporary housing units were constructed after the Great East Japan Earthquake and Tsunami (EJET),

Y. Ueda (✉)
SEEDS Asia, Kobe, Japan
e-mail: yasutaka.ueda@seedsasia.org

R. Shaw
Graduate School of Global Environmental Studies, Kyoto University, Kyoto, Japan
e-mail: shaw.rajib.5u@kyoto-u.ac.jp

R. Shaw (ed.), *Community Practices for Disaster Risk Reduction in Japan*, Disaster Risk Reduction, DOI 10.1007/978-4-431-54246-9_12, © Springer Japan 2014

Table 12.1 Damages caused by EJET (Reconstruction Agency 2013; MLIT 2013)

Human damage		Building damage			
Dead	Missing	Totally collapsed	Partially collapsed	Evacuees (3 days after the earthquake)	Temporary housing
15,883	2,671	126,458	272,191	Approx. 470,000	53,537

which occurred on March 11, 2011, and caused huge damages in Japan (Table 12.1). After the Great Hanshin-Awaji Earthquake of 1995, communities that had existed before the disaster were fragmented, as a consequence, residents of temporary housing were forced to live with non-acquaintances and in the worst case some solitary deaths happened. Therefore, to build and develop community activities for the prevention of solitary deaths was one of the challenges in temporary housing.

This chapter shows an overview of the provision of temporary housing after EJET and a case study of the community activities in temporary housing in Kesennuma City conducted by SEEDS Asia, an international non-government organization (NGO), with Kesennuma Reconstruction Association (KRA), a local community-based organization (CBO).

12.2 Situation and Challenges in Temporary Housing

12.2.1 Provision of Temporary Housing

12.2.1.1 System for the Provision of Temporary Housing Based on the Law

In Japan temporary housing is constructed by the government in accordance with the Disaster Relief Act enacted in 1947. Temporary housing is provided to *persons affected by natural disasters who lost their own house due to total collapse, complete destruction by fire, or rain and are not able to reconstruct a house by their own means*. Temporary housing is defined by law as housing *to be constructed rapidly in order for evacuees to stay safe, to promote the residents' quick transfer to their permanent house, and to be terminated by the time of their transfer* (MLIT 2012a).

Temporary housing in the law is categorized mainly into three types: (1) newly constructed temporary housing (mostly prefabricated), (2) private rental apartments, and (3) existing public housing and government-owned accommodations.

Hereafter, "temporary housing" refers to type (1) mentioned above. Approximately 40 % of the houses constructed after EJET are type (1) houses (Table 12.2).

12.2.1.2 Temporary Housing with Low Quality and Its Improvement

The law dictates that the standard size of a temporary house is should be 29.7 m^2 and construction cost of each temporary house should be less than 2,387,000 yen.

Table 12.2 Comparing each type of temporary housing (Reconstruction Agency 2012)

	Residents		Houses moved in	
Types	Number	Rate (%)	Number	Rate (%)
Temporary housing (prefabricated)	112,758	38	48,447	40
Private rental apartment	157,138	52	61,442	51
Existing public housing, etc.	30,171	10	10,824	9
Total	300,067	100	120,713	100

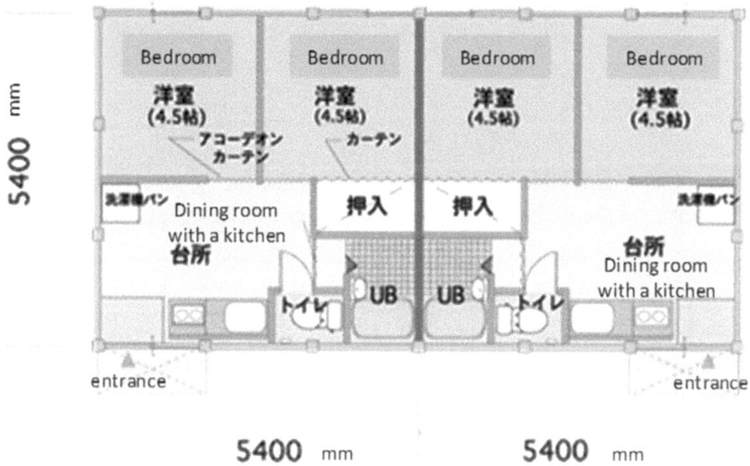

Fig. 12.1 Floor plan of the standard-size temporary housing with two bedrooms and a dining room with a kitchen (Japan Prefabricated Construction Suppliers & Manufactures Association 2013)

As a reference, the average size of a permanent house in Miyagi Prefecture, one of the affected prefectures, is 100.23 m^2 (Statics Bureau of Japan 2008). This means that the standard temporary house is one-third of the average permanent house in Miyagi (Fig. 12.1). The maximum period of using a temporary house is 2 years as a general rule. This idea was based on the concern that *temporary housing with high quality is at risk of being used as permanent housing* (Murosaki 1997). Residents of temporary housing with high quality may hesitate to move out from the housing.

These principles allow the provision of temporary housing to operate flexibly by setting special criteria for responding to disaster situations. For instance, according to the statement issued on June 21, 2011 by the Ministry of Health, Labour, and Welfare (MHLW) of Japan, additional construction cost for the improvement of residential environment such as maintenance and repair of barrier-free access, installation of insulating materials, double windows, and windbreaks were accepted by the government in the case of EJET (MHLW 2011b). As a result, the total construction cost of temporary housing reached 125,000 yen per 1 m^2 in Miyagi, approximately 1.5 times the standard stated in the law. Moreover, the period of using temporary housing is sometimes extended according to the situation of recovery from disasters. For example, extension of the period reached 5 years in the case of the Great Hanshin-Awaji Earthquake.

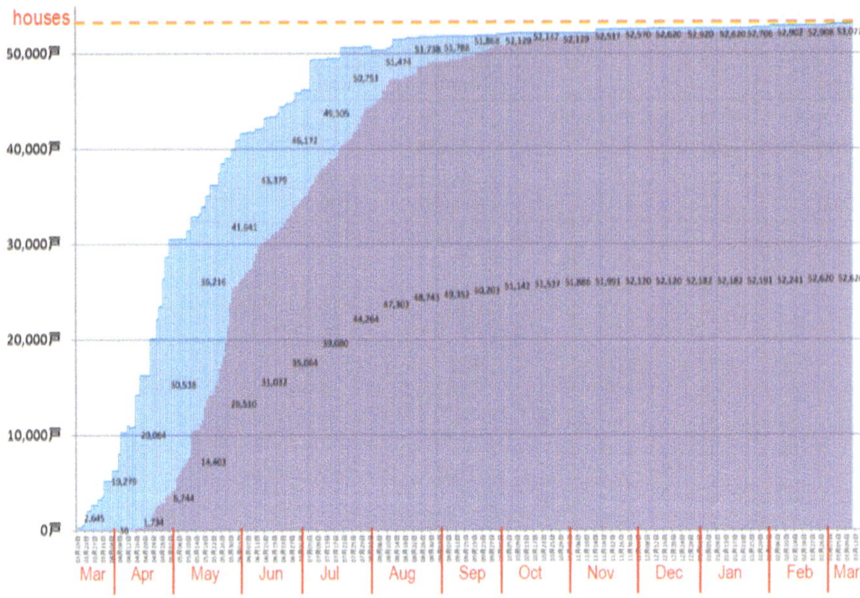

Fig. 12.2 Number of ground breaking (*blue*) and completion of temporary housing (*red*) (MLIT 2012b)

12.2.1.3 Construction of Temporary Housing

A total of 53,537 temporary housing were constructed following EJET. The first response for the construction was that of the Ministry of Land, Infrastructure, Transport, and Tourism (MLIT), which dispatched staff to affected areas on March 12, the day after the event. Construction of the housing started on March 19 (MLIT 2011). Figure 12.2 shows that the beginning of the occupancy was in April 2011.

However, half a year since the earthquake was taken to complete the construction of 50,000 temporary housing units. In the case of the Great Hanshin-Awaji Earthquake, which happened on January 17, 1995, the construction of 30,000 temporary housing units was completed in March and afterward additional order was done and finally 49,681 housing units were constructed by August 11, 1995 (Cabinet Office of Japan 1999). This means that almost half a year is required for the construction of 50,000 temporary houses in Japan. Figure 12.3 shows some typical temporary housing in Kesennuma area.

In such a scale of construction, *it was not possible for construction companies to build all the units needed immediately, due to shortages of construction materials and workers. Because of such shortages and a lack of coordination across companies, the quality and level of construction of temporary houses varies across the disaster area* (IRP et al. 2012).

In addition, the shortage of land for construction of temporary housing intensified delays of the completion. Though public spaces such as parks and grounds of

Fig. 12.3 Example of different specifications of temporary housing in Kesennuma (photos by the authors)

schools are often utilized for cost reduction in general, finding land for construction proceeded with difficulty because such public spaces were flooded by the tsunami. For example, temporary houses were prepared in a city next to Kesennuma, Miyagi Prefecture, which is located 25 km away from the affected area.

12.2.2 Features of Residents and Habitability of Temporary Housing

12.2.2.1 Method of Selection of Residents

MHLW issued a statement on April 15, 2011: "*Consideration not to break relationships of affected communities through the random selection of residents of temporary housing and not to concentrate the elderly and persons with disability at a temporary housing site should be made*" (MHLW 2011a).

However, to be fair to everyone, lottery was adopted to select residents in some municipalities. Consequently, many of them had to move with few acquaintances from the place where they lived before the earthquake.

12.2.2.2 Features of Residents

According to the survey conducted by MHLW in August 2011, the ratio of households with a single member is 15 % while that of households with two members is 32 % (Fig. 12.4 left). Furthermore, around 60 % of households have a member whose age is 65 or over and more than 15 % of households have a person with disability among its members (Fig. 12.4 right). Temporary housing is aimed to be supplied to those who lost their means. As a result most of the residents of temporary housing consist of those who need care.

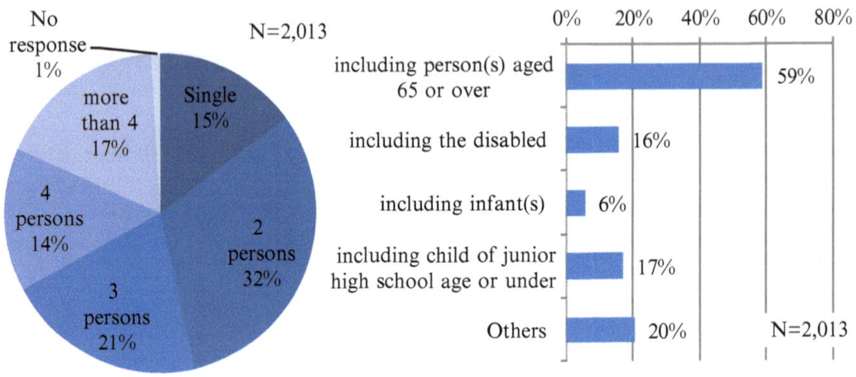

Fig. 12.4 Number of household members (*left*) and household composition (*right*) (MHLW 2011c)

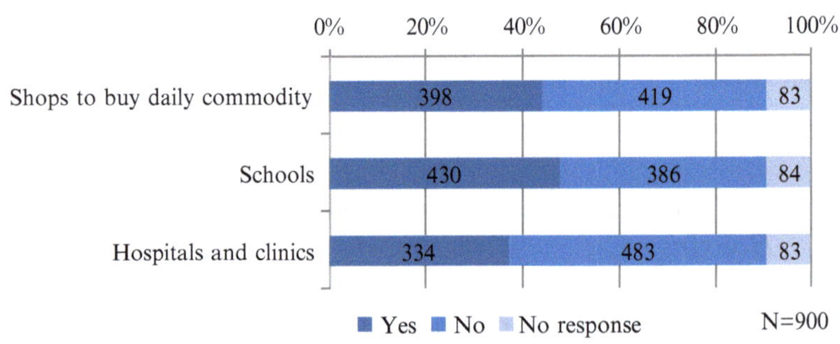

Fig. 12.5 Reachable facilities within 15 min by foot (MHLW 2011c)

12.2.2.3 Habitability

Next, the issue of location of temporary housing is focused on. According to a survey of 900 temporary housing sites conducted by MLHW, a shop to buy daily commodities, a school, a hospital, or a clinic is not located within 15 min by foot from more than half of temporary housing sites (Fig. 12.5). Moreover, the survey proved more than 40 % of the residents felt inconvenient to go shopping or visit a hospital or clinic (Fig. 12.6).

In fact, temporary housing turned out to be habitation areas that are inconveniently located and where affected people who are vulnerable and who hardly have communication with each other are gathered. Therefore, support in the welfare and health sector, such as organizing recreation activities, is needed in temporary housing.

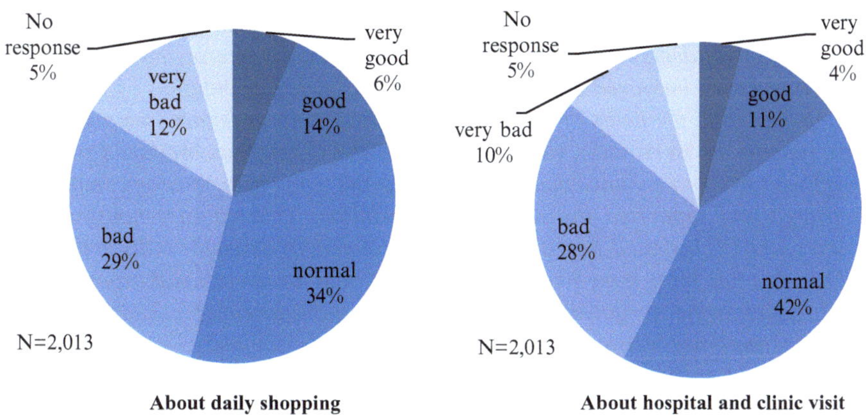

Fig. 12.6 Convenience for life in temporary housing (MHLW 2011c)

12.2.3 Challenges in Temporary Housing in Past Disasters and in This Disaster

12.2.3.1 Solitary Death in Temporary Housing

The biggest problem in temporary housing in the Great Hanshin-Awaji Earthquake was solitary death. Report by the City of Kobe (2010) is quoted below: *the city government determined it necessary to build large-scale housing complexes in the potential New Town sites in the suburban areas and the artificial islands because it was difficult to procure land in the afflicted urban areas and was necessary to build a large quantity of housing units as soon as possible. When selecting tenants, those with physical and/or mental problems such as the elderly and persons with disability were given priority for entering temporary housing units in order to extricate them from the harsh environment of local evacuation shelters as soon as possible. Since tangible and intangible support gained from human ties in local communities was an important life resource for those vulnerable people, moving them into temporary housing units meant the loss of those ties.*

Under that condition, tragic incidents, where people were found in their homes a long time after they had died, occurred. According to the Hyogo Prefectural Police Headquarters report in 1999, 233 people died alone in temporary housing in 5 years.

12.2.3.2 Problems of Support Systems for Temporary Housing in Past Disasters

The tragic incidents led the City of Kobe to begin in earnest the prevention of solitary death. According to the City of Kobe (2010), *it was found that forming a community that supplements the local ties to the communities where people previously lived is a rather effective method to unite residents in housing complexes and encourages residents to support each other through daily interaction.* And *the city government (a) improved services for the special-needs population by employing manpower from the fields of medicine, public health, and welfare (the dispatch of life-support advisers,* etc.), *(b) commissioned the Community Welfare Promotion Staff who strive to enhance mutual help within the temporary housing complex, (c) nurtured and supported the residents' associations in temporary housing complexes, and (d) installed Fureai (Interaction) Centers as assembly halls in the complexes that had more than 50 housing units and paid for operational expenses.*

In addition, *the city government promoted the establishment of residents' associations in temporary housing complexes from the time the complexes were constructed and residents started to establish residents associations in many housing sites after about three months from moving in* (City of Kobe 2010). *Therefore, newspapers started to report the establishment of residents' association in temporary housing in September 1996.* However, Tanaka (1998) indicated that *residents' association were relatively short-lived due to the absence of leaders or the frequency of changing leaders when the previous one left.* Also Shibata (1997) pointed out problems of residents' association in temporary housing, summarized as: (1) many and heavy work for board members of residents' association in temporary housing, especially the responsibility on privacy and life-threatening matters and (2) a lot of trouble regarding the management of subsidies from the municipality for the residents' association. As a result, a network among representatives of residents' associations was formed to exchange opinions regarding the problems and to try to find solutions.

12.2.3.3 Countermeasures in EJET

Given such lessons learned from the Great Hanshin-Awaji Earthquake, measures for community building in temporary housing were taken as below:

(a) **Measures in layout planning**: a common room was set in sites of temporary housing where houses were more than 50 in accordance with the policy by the government. Besides, layout planning in consideration of promoting communication, such as having entrances facing each other or having temporary shops, was adopted in some temporary housing sites (Fig. 12.7).
(b) **Measures in the selection of residents**: In the case of Nagahora Area in Rikuzentakata City, Iwate Prefecture, affected community members negotiated with land owners in the village where they had lived to secure land for the construction of temporary housing. Afterward, they convinced the municipality to construct 26 temporary housing units so that they could live together. This is a rare case that the relationship in the affected community was kept in the temporary housing (Nagahoragenkimura 2013).
(c) **Enhancement of care activities**: According to the Japan National Council of Social Welfare (2012), the number of Life Support Advisers (LSAs) was increased after EJET as well as in past disasters. The total number was 567 in

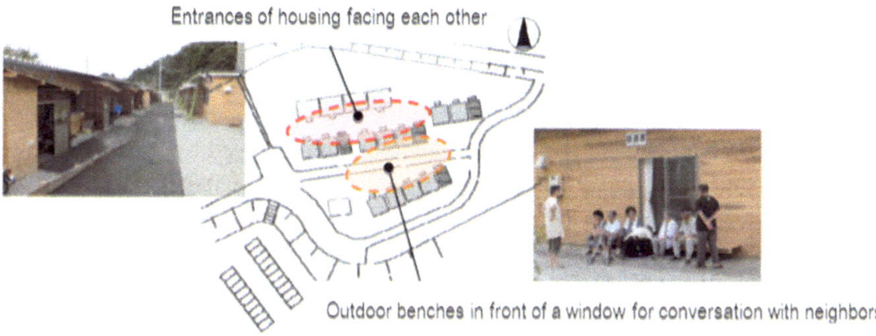

Fig. 12.7 Layout planning for promoting communication in Miharu-cho, Fukushima (MLIT 2012a)

61 municipalities as of March 1, 2012. Also, encouragement of exchange activities in temporary housing, such as KRA activities to be mentioned later, was conducted in some areas.

(d) **Promoting the establishment of residents' associations**: municipalities in the affected areas held explanation sessions and promoted to establish residents' associations for sharing information on problems regarding residents in temporary housing and to find solutions by themselves. MHLW (2012) reported 88.9 % of temporary housing sites had a residents' association as of March 9, 2012, around 1 year after the event. This number is higher than the 63.1 % in the case of the Great Hanshin-Awaji Earthquake.

As mentioned above, there are some cases for consideration on layout planning or selection of residents in EJET. However, many cases actually involved a uniform layout and the selection of residents broke the relationships in the affected community made before the disaster. For this reason, (c) enhancement care activities and (d) promoting establishment and development of residents' association are vital in this disaster.

Case studies of these important two activities are given to the next section.

12.3 Community Activities in Temporary Housing: Case of Kesennuma City

12.3.1 Overview of Damages in Kesennuma City

Kesennuma City is located in the north east end of Miyagi Prefecture, along the coast facing the Pacific Ocean. It has an area of 333 km^2. The main industries of the city were fisheries and tourism. The city has a well-developed a seafood-processing industry from early times. The current Kesennuma City comprises of three administrative districts: the former Kesennuma City, Karakuwa Town, and Motoyoshi Town.

After completion of the consolidation of these three municipalities in September 2009, one administrative office is placed in each area of the three districts.

The city was devastated by EJET. The tragedy of the city is that in addition to the damage caused by the tsunami, the central part of the city where many seafood processing companies were located, was completely destroyed by the massive fire that spread from an oil tank located at the bay entrance. According to the city office, 1,280 died or were missing, with 15,751 losing their homes, affecting 9,500 households (Kesennuma City 2013). The city had 26,622 households in 2010. This means that more than one-third of the population were affected. Also, out of 4,102 registered business institutions, 3,314 businesses or 80.8 % of the total were considered to be affected and the local economy was devastated, leading to a huge number of unemployed people (Kesennuma City 2011).

12.3.2 Situation and Problems in Temporary Housing in the City

Ninety-three sites consisting of 3,503 temporary housing units were prepared by the city. Construction of the sites went on all over the city until the end of December 2012 due to lack of construction sites. The last unit was completed 8 months after the completion of the first one in April 2011. Various sizes of temporary housing sites consisting of a minimum of 5 to a maximum of 228 houses were constructed. As each site was completed, people moved in. To be fair to everyone, residents were selected by lottery. Consequently, many of them had to move from the place where they lived before the earthquake to the site with non-acquaintances. Support for community activities to prevent solitary death was needed in light of lessons learned from the Great Hanshin-Awaji Earthquake.

12.3.3 Case 1: Tea Parties for Community Development in Temporary Housings

First, the case of tea parties held by KRA in cooperation with SEEDS Asia is picked up. KRA was established in April 2011 for the purpose of supporting work on the reconstruction of the affected area by the affected people themselves. KRA started an intergenerational exchange project as the contractor of Kesennuma City office since June 2011. The main activity of this project became the tea parties covering 90 temporary housing sites (all sites in the city except three for group home), acting on the technical advice of SEEDS Asia.

12.3.3.1 Overview of Tea Parties

Because of the increase of temporary housing, KRA recruited new staff members to enable them to have tea parties at least once a month at each temporary housing site. From the official start on July 2011 to March 2012, tea parties had been conducted cumulatively 431 times.

Fig. 12.8 Tea parties held outdoors (*left*) and indoors (*right*) (photos by the authors)

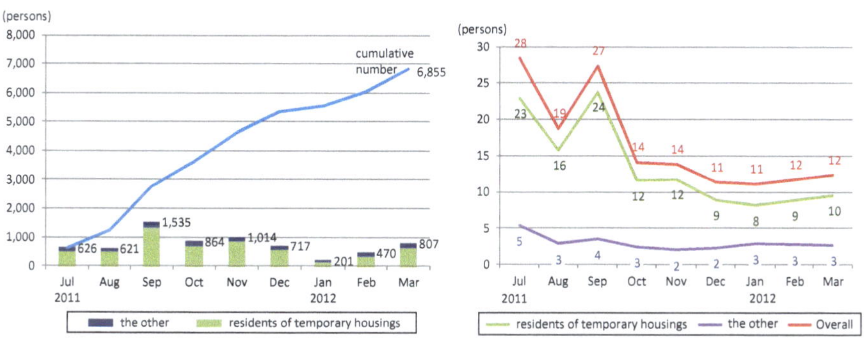

Fig. 12.9 Number of participants in tea parties (*left*) and the monthly average (*right*)

One of the objectives of tea parties is for residents to build relationships and to know each other, which hopefully would lead to the establishment of a residents' association. For example, during conversation in tea parties, KRA staff would explain that "if someone keeps the keys to common rooms in the site, it makes it easier to hold a tea party." "Residents' associations will help aid support organizations to operate more easily." After the establishment of the association, arrangement of tea parties was done through the representatives of the association and tea parties shifted to be held outdoors to the common rooms inside temporary housing sites (Fig. 12.8).

12.3.3.2 Decreasing Number of Participants in Tea Parties

The number of participants in tea parties is cumulatively 6,855 from July 2011 to March 2012 (Fig. 12.9 left). The monthly average number of participants remained at a high level from July to October. However, starting from November, the number decreased and became around ten (Fig. 12.9 right).

Tea parties have two or three participants other than the residents of temporary housings on average per month. Most of them were staff of external support

Table 12.3 Recreational events proposed during a project meeting (SEEDS Asia 2012)

Food	Handcraft	Skill-learning	Hobbies
• Soba noodle making • Udon noodle making • Cooking for beginners • How to cook fish • Cooking ethnic food • Sweets making • Baked potatoes • Stewed potato party • Rice cake making	• Hammock making • Ornament making • Bracelet making • Carp streamer making • Kite making • Takeuma (Stilt) making • Beanbags making • Map/Model making • Staffed animal making • Greeting card making • Business card making • Knitting	• Personal computer • Foreign language • Usage of electronics • Usage of mobile phone • Book keeping • Carpentry • Calculation • Kanji • Crime prevention • Disaster risk reduction	• Watching movies • Haiku/Tanka • Waka poems • Dialect poems • Calligraphy • Card game • Japanese chess • Pottery • Indian ink painting • Printing • Drawing • Bonsai • Ventriloquism • Karaoke
Exercises	*Fashion/Beauty*	*Outdoor Activities*	*Exchange*
• Walking • Baseball • Stretching • Yoga stretching • Aerobics • Hula • Social dance • Kendo game	• Clothing coordination • Make-up • Nail-care • Hair-cut	• Vegetable farming • Fishing • Picking mountain vegetables • Birds watching • Cross-country skiing • Shiitake mushroom farming	• Children events • Christmas party • Animal therapy

groups such as LSA, NGO staff, or volunteers. That is to say, communication opportunities given by tea parties functioned as windows for matching with external support.

12.3.3.3 Recreational Events in Tea Parties

In order to encourage people who stayed in temporary housing and were not interacting with other people to come outside for interaction, recreational events were proposed (Table 12.3). Planned recreational events are familiar to the residents which helped to promote communication among the residents.

12.3.3.4 Medical Care and Welfare Activities in Tea Parties

Regarding the number of additional events in tea parties, craft making which leads to occupational therapy was the highest in December while medical and health check-up was the highest in January and February (Fig. 12.10 left). The reason was that there was a request to receive medical specialists from residents of temporary housing in order for them to keep their health. Figure 12.10 (right) shows more medical and welfare specialists joined the tea parties. Thus, it is found that tea parties were opportunities to respond to needs in the field of medical care and welfare in temporary housings.

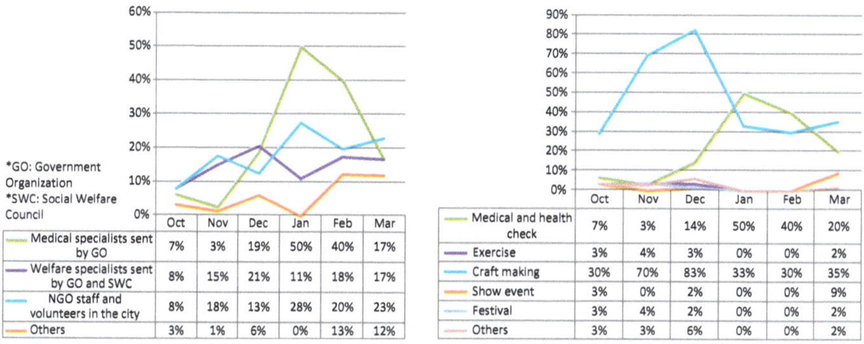

Fig. 12.10 Rate of additional events (*left*) and rate of external participants (*right*)

Table 12.4 Monthly average number of participants per tea party with additional events

	2011						2012		
Items	July	August	September	October	November	December	January	February	March
Overall	28	19	27	14	14	11	11	12	12
Medical and health check	–	–	–	8	26	15	12	14	16
Exercise	–	–	–	13	21	17	–	–	28
Craft making	–	–	–	11	11	10	13	10	13
Show event	–	–	–	8	–	6	–	–	29
Festival	–	–	306	125	83	23	–	–	31
Others	–	–	–	22	23	18	–	–	12

12.3.3.5 Participation of Children for Intergenerational Exchange

Regarding the tea parties with additional events, some of them have very high participation, especially the festival in September with 306 participants and the festival in October with 125 (Table 12.4). Children participation is considered as one of the reasons. Normally, recreation for children is organized during festivals. Then family members come to the event with their children. In such situations, communication among residents is accelerated. Thus, children participation becomes a key for community activities.

12.3.3.6 "Promotion of Communication" to "Coordination with External Support"

Communication opportunities for residents in temporary housing can be utilized to understand their needs and to provide services in response to these needs. In addition, sharing information on measures to address the needs is important to enhance

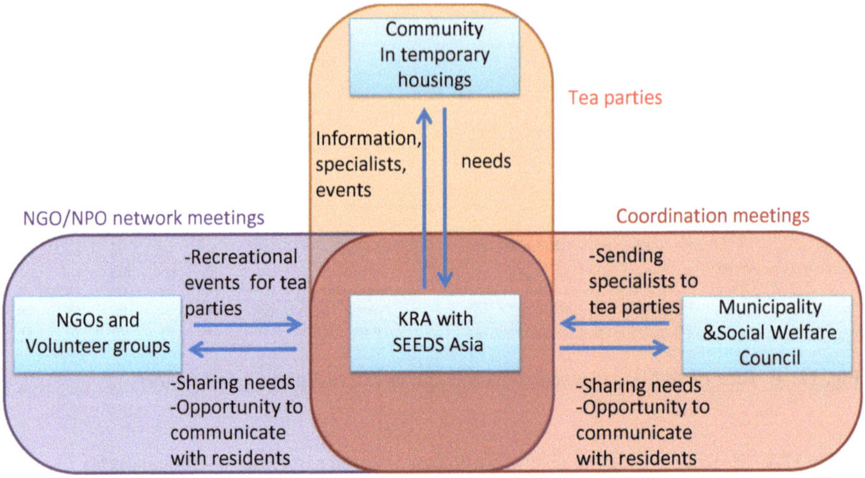

Fig. 12.11 Diagram showing KRA activities and other actors (*right*)

community activities. Therefore, KRA and SEEDS Asia began to make a community paper which included administrative and support information and to distribute it to the temporary housing sites in September 2012.

Furthermore, KRA got engaged in NGO/NPO network meetings and coordination meetings with officials and staff of social welfare council for medical, health, and welfare sectors. Thus, the tea parties organized by KRA have played a role in the promotion of communication as well as in providing an intermediate organization to act as bridge between residents and external support such as officials, specialists, and NGOs (Fig. 12.11).

12.3.4 Case 2: Exchange Meetings Among Representatives of Temporary Housing

Another case is the exchange meetings among representatives of temporary housing sites in Kesennuma Area (former Kesennuma City area) held by SEEDS Asia, as an occasion to figure out solutions together to problems at each site.

12.3.4.1 Process of the Exchange Meeting

The exchange meetings was triggered by a discussion session organized by SEEDS Asia in December 2011 with a practitioner who have had watching activities for affected people living in temporary housing and public housing after the Great Hanshin-Awaji Earthquake. This session was attended by representatives from eight residents' associations of temporary housing. Winter is severe in Kesennuma and there were concerns on how to cope with the cold in temporary housing sites. Some representatives of residents' associations individually requested the city for possible

1st meeting: January 28, 2012, 22 representatives participated

- Establishment of an association called "Kesennuma Temporary Housing Network."
- Submission of collective request to the city office swiftly after problems were brought up at meetings.
- Dividing Kesennuma area into smaller parts and selecting a representative for each part.
- Sharing results with absentees of the meeting, and encouraging future attendance.
- Aid organizations support residents' associations to hold area meetings to consolidate their needs.

Follow-up of the 1st meeting: February 7, 2012, 11 representatives participated

- Review the requests to city office as many of them were solved by residents or with the support of aid organizations and only those that needed governmental support would be submitted.
- Tentative chairperson of "Kesennuma Temporary Housing Network," who was a leader of a residents' association and voluntarily worked on various issues of temporary housing, was selected.

2nd meeting March 15, 2012, 12 representatives participated

- Future exchange meetings will be continued by a local CBO, whose representative is the tentative chairperson of "Kesennuma Temporary Housing Network".

Fig. 12.12 Process of the meetings among residents' association representatives

measures, but no collective action had been taken. Through the session, all participants agreed to establish a working group on common problems to make requests to the city for solution and to have another exchange meeting for all representatives in the Kesennuma Area to proceed with the matter. Afterward, the exchange meeting was held in coordination with SEEDS Asia. Representatives of residents' association from 59 sites were called for. The process is shown in Fig. 12.12.

12.3.4.2 Focus on the Process of Identifying and Solving Problems by the Representatives

Now, the process of sharing problems and finding solutions by the representatives is focused on (Fig. 12.13). Twenty-seven problems were identified in the first meeting. Then the problems were divided into ones to be solved by residents themselves and ones that could not to be solved. Furthermore, tentative chairperson who had been chosen in the follow-up meeting visited the city to hear a response to seven structural problems which hadn't found a solution. The result was reported in the second meeting. Though the answer from the city for the seven problems was that it was hard to come up with measures due to budget limitation, holding continuous exchange meetings was approved in the second meeting with the reason that joint activities among the representatives for sharing common problems and finding solutions was important.

Fig. 12.13 Consensus of meetings among residents' association representatives

12.3.4.3 Continuation of the Exchange Meetings by a Representative of Temporary Housing

Another point is that the continuous meetings would be organized by a local CBO which had been founded by the tentative chairperson in 2012. Consequently, this exchange meeting which had been started by SEEDS Asia, one of the external organizations, as a direct support project, had been shifted to become a self-reliant project by representatives concerned in a process involving a participatory approach.

Since then, the meeting has been run by the local CBO with following cycle from April of 2012: (1) organize a meeting, (2) make sure the problems are clearly understood in the meeting, (3) come up with action to solve the problems, such as discussion with the city or holding a workshop on the solution, and (4) organize a meeting again to report the result.

12.4 Key Findings and Lessons

Key findings and lessons for community development in temporary housing analyzed through two cases are summarized below:

(1) The process of re-discovering and understanding problems, considering and responding to the problems, and sharing information about the matter by providing communication opportunities is important for community building.
(2) The relationship with the city government, specialists, and coordinators for making communication opportunities is needed for this process. Such communication opportunities give a chance to know the needs of and to provide services to residents.

(3) The concern of residents in the field of medical and health care and welfare is enlarged in temporary housing. In addition, requests to improve structural problems are increased due to the poor quality of the houses. Therefore, it is important for community development in temporary housing to organize a process that allows residents to take action through the enhancement of consciousness of the residents on the issues.
(4) Children participation is a key to activate the community. Relationship with schools and/or kids club under community centers should be made to promote children participation in community activities in temporary housing.

Temporary housing sites where residents were selected by lottery such as the ones in this case study are an epitome of a vulnerable aged society so that the above lessons are significant not only for temporary housing but also for existing communities.

Acknowledgement The case study in this chapter is compiled on the basis of findings and lessons from community recovery projects conducted by SEEDS Asia with support from Kesennuma Reconstruction Association and Church World Service (CWS). The authors are grateful to them.

References

Cabinet Office (1999) Records of reconstruction from the Great Hanshin-Awaji Earthquake (Hanshin-Awaji Daishinsai fukko shi). http://www.bousai.go.jp/kyoiku/kyokun/hanshin_awaji.html. Accessed 27 July 2013 (in Japanese)

City of Kobe (2010) Comprehensive strategy for recovery from the Great Hanshin-Awaji Earthquake. http://www.city.kobe.lg.jp/safety/hanshinawaji/revival/promote/img/English.pdf. Accessed 27 July 2013

IRP (International Recovery Platform) et al (2012) Knowledge note 4 transitional shelter, the Great East Japan Earthquake learning from megadisasters, The World Bank. http://wbi.worldbank.org/wbi/Data/wbi/wbicms/files/drupal-acquia/wbi/drm_kn4-3.pdf. Accessed 27 July 2013

Japan National Council of Social Welfare (2012) Report on current situation and challenges of life support and consultation to affected people by Social Welfare Councils in affected area by EJET (Higashinihondaishinsai hisaichisyakyo niokeru hisaisya heno seikatsusaiken/soudankatsudo no genjo to kadai). http://www.shakyo.or.jp/research/11support.html. Accessed 27 July 2013 (in Japanese)

Japan Prefabricated Construction Suppliers & Manufactures Association (2013) Japan Prefabricated Construction Suppliers & Manufactures Association website. http://www.purekyo.or.jp/. Accessed 27 July 2013 (in Japanese)

Kesennuma City (2011) Kesennuma City reconstruction plan (*Kesennuma-shi Fukko Keikaku*). http://www.city.kesennuma.lg.jp/www/contents/1318004527115/files/hukkokeikaku.pdf. Accessed 27 July 2013 (in Japanese)

Kesennuma City (2013) Damages in Kesennuma (Kesennuma-shi no higai jokyo) as of 28th February 2013. http://www.city.kesennuma.lg.jp/www/contents/1300452011135/index.html. Accessed 27 July 2013 (in Japanese)

MHLW (Ministry of Health, Labour and Welfare) (2011a) About temporary housings in EJET, vol 1 (*Higashinihondaishinsai nikakaru Oukyukasetsujutaku nitsuite sono 1*) issued on 15th April. http://www.mhlw.go.jp/stf/houdou/2r98520000019dbz-img/2r98520000019izs.pdf. Accessed 27 July 2013 (in Japanese)

MHLW (2011b) About temporary housings in EJET, vol 3 (*Higashinihondaishinsai nikakaru Oukyukasetsujutaku nitsuite sono 3*) issued on 21st June. http://www.mhlw.go.jp/stf/houdou/2r9852000001g837-att/2r9852000001g8ls.pdf. Accessed 27 July 2013 (in Japanese)
MHLW (2011c) Report of survey on residential environment in temporary housings (*Oukyukasetsujutaku no kyojukankyo nado ni kansuru anketo chosa*). http://www.mhlw.go.jp/stf/houdou/2r9852000001pw1l-att/2r9852000001pw63.pdf. Accessed 27 July 2013 (in Japanese)
MHLW (2012) Report on establishment of residents' association in temporary housings (*Okyukasetsujutaku niokeru jitikai no secchijoukyou nit suite*) issued on 15th 2012. http://www.mhlw.go.jp/stf/shingi/2r985200000256wr-att/2r98520000025 70k.pdf. Accessed 27 July 2013 (in Japanese)
MLIT (Ministry of Land, Infrastructure, Transport, and Tourism) (2011) Materials of kickoff meeting on promotion of supplying temporary housings to affected people (*Hisaisya muke no jutakukyokyu no sokushin nado ni kansuru kentokaigi daiikkai siryou*) issued on 28th March. http://www.mlit.go.jp/jutakukentiku/house/jutakukentiku_house_fr4_000010.html. Accessed 27 July 2013 (in Japanese)
MLIT (2012a) Handbook on construction of temporary housings – interim report (Oukyukasetsujutakukensetsuhikkei – Chukantorimatome). http://www.mlit.go.jp/common/000211741.pdf. Accessed 27 July 2013 (in Japanese)
MLIT (2012b) Shift of number of groundbreaking and completion of temporary housings (Okyukasetsujutaku chakko kansei kosu no suii) issued on 11th March 2012. http://www.mlit.go.jp/common/000143900.pdf. Accessed 27 July 2013 (in Japanese)
MLIT (2013) Situation of groundbreaking and completion of temporary housings (Okyukasetsujutaku chakko kansei jokyo) issued on 1st April 2013. http://www.mlit.go.jp/common/000140307.pdf. Accessed 27 July 2013 (in Japanese)
Murosaki Y (1997) Construction of temporary housings and problems on living in the housings (Kasetsujutaku no kensetsu to seikatsujo no mondaiten). In: Life of affected people with struggle – Series of researches on the Great Hanshin-Awaji Earthquake, vol 2 (Hanshindaishinsai kenkyu 2 – Kuto no hisai seikatsu), Kobe Shinbun Sogo Syuppan Center, pp 115–128 (in Japanese)
Nagahoragenkimura (2013) Nagahoragenkimura website. http://www.nagahoragenki.jp/. Accessed 27 July 2013 (in Japanese)
Reconstruction Agency (2012) Situation and measures of reconstruction (*Fuko no Genjo to Torikumi*) issued on 14th December. http://www.reconstruction.go.jp/topics/20121225_sanko03.pdf. Accessed 27 July 2013 (in Japanese)
Reconstruction Agency (2013) Situation and measures of reconstruction (*Fuko no Genjo to Torikumi*) issued on 2nd July. http://www.reconstruction.go.jp/topics/main-cat7/sub-cat7-1/20130704_sanko01.pdf. Accessed 27 July 2013 (in Japanese)
SEEDS Asia (2012) Community recovery of the Great East Japan Earthquake and Tsunami - Kesennuma experiences. Publication, SEEDS Asia
Shibata K (1997) Current situation of resident's association's activities in temporary housing (Kasetsujutaku niokeru jichikaikatudo no genjo). In: Life of affected people with struggle – Series of researches on the Great Hanshin-Awaji Earthquake, vol 2 (Hanshindaishinsai kenkyu 2 – Kuto no hisai seikatsu), Kobe Shinbun Sogo Syuppan Center, pp 129–137 (in Japanese)
Statics Bureau of Japan (2008) Housing and land survey. http://www.stat.go.jp/english/data/jyutaku/index.htm. Accessed 27 July 2013
Tanaka M (1998) Study on management of temporary housings (*Kasetsujutakuunei ni kansuru kenkyu*). A comprehensive study of the management of emergency shelter systems after a large scale disaster (*Daikibosaigai ni okeru hinansoshikiunei ni kansuru sougoutekikenkyu*), Saigaienjokenkyukai. http://www.syaanken.or.jp/?p=2742. Accessed 27 July 2013 (in Japanese)

Chapter 13
Adapting Fisheries to Climate Change: Community-Based and Scientific-Oriented Fisheries Management of Saroma Lake, Hokkaido

Shimpei Iwasaki

Abstract This chapter presents a case study of Saroma Lake fisheries by highlighting adaptive responses to risk reduction coupled with climate change. Saroma Lake is the southernmost area of seasonal sea ice distribution in the northern hemisphere. Changes in temperature are likely to impact negatively on the ice-covered ecosystem and fishing production in the lake where the fishers need to adapt to the impacts. The case study reveals that innovative efforts performed by the fishers and researchers have been made by applying community-based and scientific-oriented fisheries management. The fishers united together and built an integrated body, leading them to achieve cooperative fishery governance system. The system not only ensures a holistic approach to respond to changes in the lake ecosystem, but also reduces a range of risks through active use of scientific knowledge by employing full-time researchers in their structure as well as strengthening interactions between fishers and outside researchers. The employed researchers work on problems of interest to the fishers and search workable the solutions. Permanent or long-term residency can allow them to interpret concerns of the fishers and act in a leading role in the coordination of adaptive fisheries development. Based on the Saroma Lake experience, this chapter will provide potential initiatives to help develop a pathway for linking fisheries management to risk reduction.

Keywords Adaptive management • Climate change • Fisheries • Resident researchers • Saroma Lake

---

S. Iwasaki (✉)
Department of Environmental Sciences, Fukuoka Women's University, Fukuoka, Japan
e-mail: shimpei.iwasaki@gmail.com

R. Shaw (ed.), *Community Practices for Disaster Risk Reduction in Japan*, Disaster Risk Reduction, DOI 10.1007/978-4-431-54246-9_13, © Springer Japan 2014

13.1 Introduction

Existing fisheries management poses challenges for ensuring a wise use of fishery resources. Fishery resources are becoming fully exploited or overexploited all over the world (FAO 2012). In Japan, fish landings in 2011 (approximately 4.77 million tons) decreased to a peak of 63 % of its peak in 1984 (Fisheries Agency 2013). The decline in the catch is greatly linked to fishing activities, but recently, there has also been a growing concern over the impacts of climate change for fish production and the state of aquatic ecosystems (Allison et al. 2009; Brander 2010; Katsuki et al. 2012; Rijnsdorp et al. 2009). Climate-induced impacts on fisheries are due to a variety of direct and indirect effects of multifaceted factors including temperature, winds, vertical mixing, salinity, oxygen, pH and others (Brander 2010). Out of them, changes in temperature, even small changes in water temperature are expected to exert strong pressure upon fish ecology (WWF 2005). Given warming atmosphere of the climate system is now unequivocal (IPCC 2007), fishers and relevant stakeholders need to adapt to the impacts from changes that are already unavoidable (UNISDR 2007).

Changes in temperature vary regionally, but greater warming is expected to occur at high latitudes (Anthony et al. 2009). Surface and satellite-based observations indicate a substantial decrease of northern hemisphere sea ice particularly in the Arctic sea during the past few decades (Perovich and Richter-Menge 2009; Stroeve et al. 2011; Vinnikov et al. 1999). Although impacts of reduced sea ice extent do not yet fully understood, these are expected to include increased autumn and winter temperatures, stronger wave activity, intensified coastal erosion, disrupted thermohaline circulation (Arrigo et al. 2008). Climate risk may change habitats for animals, affecting people's ability to harvest resources such as the fisheries and hunting (Ford et al. 2009).

Provided greater uncertainties and regional variability inherent in fisheries management, an adoption of command-and-control measures makes it difficult to achieve a wise use of fishery resources. To cope with it, many scholars increasingly put high emphasis on the significance of the adaptive management approach (Brander 2010; Armitage et al. 2007; Berkes 2004; Berkes 2005; Carlsson and Berkes 2005). Adaptive management as an approach takes into account uncertainty and assumes that management knowledge is inadequate, thus requiring experimentation and learning in an iterative process (Berkes 2007). With this recognition, this chapter presents a case study of Saroma Lake by highlighting adaptive responses to changes in the fisheries. Saroma Lake is located in the northeast of Hokkaido along the Okhotsk sea which is the southernmost area of seasonal sea ice distribution in the northern hemisphere (Ohshima et al. 2000). During the period from 1977 to 2011, average annual maximum and minimum temperatures in the Saroma Lake region are 32.7 and −24.2 °C, respectively (Japan Metrological Agency 2013). Seemingly, there is a slight warming trend of maximum and minimum temperature especially in the winter season (Iwasaki and Shaw 2010). The warming trend of temperature affects duration of complete ice coverage in Saroma Lake, which may affect adverse impacts on aquatic animals.

The vulnerabilities of those who are dependent on climate sensitive aquatic resources and need to build their resilience to cope with changes in temperature are among the more important challenges in Saroma Lake.

From the next section, this chapter delineates an evolutionary perspective of adaptive fisheries management to respond to risk reduction in Saroma Lake in the context of climate change. Firstly, the research describes a profile of the case study site. Secondly, it explores performances done by fishers and relevant stakeholders in responding to environmental and climate changes. Of particular note is community-based and scientific oriented management in adapting the fisheries to changing marine ecosystem by introducing a research department (resident researchers[1]) in their management structure. Based on the Saroma Lake experience, this chapter will provide potential initiatives to help develop a pathway for linking fisheries management to risk reduction.

13.2 Profile of Case Study Site (Saroma Lake)

Saroma Lake is the largest coastal lagoon in Japan, situated at latitude 44°05′07″ and 44°11′58″ north and longitude 143°40′06″ and 143°58′14″ east (see Fig. 13.1). The lake has semi-closed estuaries with sea mouths between Okhotsk sea and lake. In the lake, two artificial sea mouths have been excavated where the water exchange can be maintained. Around 90 % of the total inflow from the sea to the lake passes through the former mouth which was opened in 1927. The salinity level in Saroma Lake is almost similar to that of the Okhotsk sea due to the active tidal water exchange through the two mouths. The lake receives fresh water from 13 rivers, especially two principal streams (i.e., River Saromabetsu and Baro) where a large quantity of freshwater and subsequent sediments and nutrients are supplied into the lake.

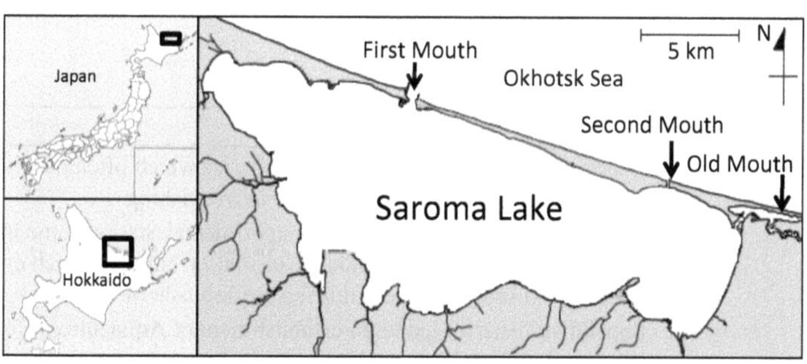

Fig. 13.1 Map of Saroma Lake

[1] The term resident researchers refer to persons who promote the trans-disciplinary solution-oriented research by committing themselves as stakeholder in the local community (ILEK 2013).

The lake is typically subarctic climate and characterized by pelagic and ice-covered ecosystem. High water temperatures and high salinity levels observed from May to October derive from the Soya warm current, which is dominant in summer season. By contrast, low temperatures and low salinity levels observed from November to April drive from the East Sakhalin cold current, which is dominant in winter season (Sato et al. 2007). In winter, the lake becomes totally or partially ice coverage while drift ice, which starts forming at points where the Amur River to the north in Sakhalin flows into the sea of Okhotsk, surges into the lake through the channels.

The sea of Okhotsk is considered one of the richest north temperate marine ecosystems in the world. In spring, the Okhotsk sea receives a large volume of freshwater and nutrients from rivers commensurate with ice melting, leading to phytoplankton blooms and the subsequent primary production in the region. The coastal lagoon environment connecting to the Okhotsk sea endows it with a highly productive environment in which fishing is a principal benefit. Particularly, Saroma Lake is known as one of the most prominent scallop production places in Japan. More than half of the water area in the lake is utilized for scallop culture. The annual yield of breeding and culturing scallop fisheries is approximately 57,987 tons in 2007, accounting for around 79 % of the total fish landings in the covered three villages (marine scallops—68.5 %, lake scallops—8.5 %, juvenile scallops—2.0 %).

13.3 Development Process of Scallop Culture in Saroma Lake

This section sets out to explore a historical discussion on fishery development in Saroma Lake. As above mentioned, breeding and culturing scallops are fundamental for the lake fisheries, contributing to a great extent to stable aquatic production in the region. The fishing operation was a result of iterative learning process by dedicated fishers and researchers (Table 13.1).

Table 13.1 Development process of scallop culture in Saroma Lake

Period	Events
Before 1928	Fishery operation was based on capture fisheries which often made fishery resources depleted or deteriorated due to overfishing
1931	An engineer from Hokkaido Fisheries Experimental Station came in Saroma Lake and discovered the potentiality of scallop culture, which enabled fishers to start breeding and culturing scallop fisheries
1952	The cooperative fisheries led to an establishment of Aquaculture Cooperation of Saroma Lake (ACSL), leading to build a nested cooperative partnership among three fishery cooperative associations
1961 to Present	Elaborations were made on improvement of fishing methods and cultivation of large-scale scallops by the initiatives of fishers and researchers. Having recognized the significance of scientific-oriented management, a research department under ACSL was set up in 1987, to respond to changes in the fisheries

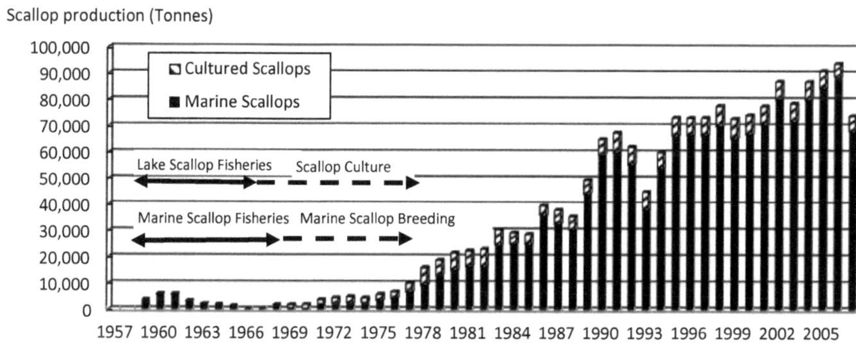

Fig. 13.2 Trend of scallop production in the Saroma Lake region (modified from ACSL data)

Interactions between researchers and fishers became the turning point to develop scallop culture. In 1931, a researcher from Hokkaido Fisheries Experimental Stations visited the lake with an aim to conduct a survey related to revitalization of oyster population. The researcher happened to discover a lot of juvenile scallops on the surface of oyster shells when collecting oyster seedlings. Having recognized the significance of the possibility of scallop culture in the lake, the researcher further strived to conduct experimentation of scallop culture and manage to get the business for the fishers on its feet. Since then, the fishers have interacted closely with researchers working on development of scallop culture and problems of the interest to the fishers (ACSL 1999a). Indeed, 1 month training for required knowledge and skills of scallop culture was provided to 20 active young fishers. Most of the participants played a leading role in developing and managing breeding and culturing scallop fisheries in Saroma Lake, some of whom became leaders of three fishery cooperative associations (FCAs).

The cooperative fisheries strengthened their fishery governance system among three FCAs by establishing integrated body so-called Aquaculture Cooperation of Saroma Lake (ACSL) in 1952. ACSL served as coordination of affiliated three FCAs, enabling to share their fishing grounds as common property resources in the lake beyond the border line of administrative territory. After a series of attempts done by the fishers and researchers, stable production of scallop culture has been developed since 1970s (Fig. 13.2).

In the development process of scallop culture, it is important to note that eutrophication along population growth, rapid industrialization and scallop culture affected adverse impacts on aquatic resources in the lake, some parts of where red tide has taken place in 1980, 1983 and 1993, respectively. In response to this, there occurred a massive death of cultured scallops in Wakasato area of Saroma village for the first time in 1976. The accident has been instantly expanded to one after another, leading to almost the entire area in the lake (ACSL 1999b). Related to this, the production of 3-year cultured scallops has been decreased up to 3,673 tons in 1984, compared with 6,728 tons in 1980. Recognizing such negative impacts, the fishers contacted closely with relevant researchers from outside and applied ecosystem approach in a way that cultured scallop allowance limit (CSAL) commensurate with environmental carrying capacity in the lake was introduced. That is, the fishers requested them to carry out surveys on setting environmental carrying capacity in Saroma Lake and its evaluation of current fisheries management. Then, their findings revealed that production of cultured scallops be 50 million species at maximum: the fishers must reduce their production up to –27 %. On account of this, the fishers implemented necessary monitoring with strong leaderships to successfully keep their production below the recommendation under the banner of cooperative fishery governance through ACSL, contributing to resolution of massive death of cultured scallops in the lake in 1985.

In this way, the fishers in Saroma Lake responded to the challenge in the development process of breeding and culturing scallop fisheries, by building good partnership between the fishers and researchers as well as promoting cooperative fisheries management among the members. In the former, the fishers requested researchers to develop a better understanding of environmental factors on how negative impacts are influenced in resource management. Throughout the linkage, the fishers became aware of the necessity of scientific-oriented decisions to adapt to changes in the changing ecological-social-economic system. Having recognized the significance of scientific-oriented management, a research department under ACSL was set up in 1987. As a consequence, ACSL funded out of their profits mostly from scallop culture and employed full-time researchers who are expected to make the long-term commitments that are required from the fishers.

13.4 Emerging Issues of Global Warming in Saroma Lake Fisheries

Apart from accelerated anthropogenic pressures from the upstream and scallop culture in the lake environment, it has been reported that the fisheries require coping with warming atmosphere of the climate system (Iwasaki and Shaw 2010; Shiomoto 2008; Shiomoto et al. 2012). There is a slight warming trend of maximum and minimum temperature especially in winter season as the above mentioned. Changes in temperature, even small changes in water temperature are expected to exert strong pressure upon fish ecology (WWF 2005). It is reported that although the peak scallop production was September to November during 1970s, a massive death of cultured scallops has been taken place in summer due to higher temperature, requiring shipping adjustment by August in some fishing grounds (SCECSL 2005).

Furthermore, it is highly possible that a trend toward shortening of the duration of ice coverage has been reported in the lake due to the increased temperature. Thinning of the ice coverage has also tended to be shorter. The warming trend can possibly change ice-covered ecosystem and aquatic biomass in the fisheries during winter and spring as follows.

First, the ice coverage in Saroma Lake has a tendency to shorten the period in winter (Fig. 13.3). The same trend is being reported globally for ice-covered aquatic

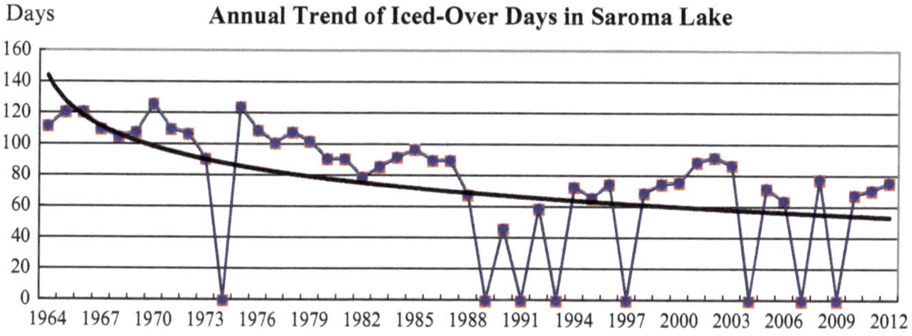

Fig. 13.3 Shorter period of ice distribution in Saroma Lake (modified from ACSL data)

areas (Shiomoto et al. 2012). In Saroma Lake, it needs to be mentioned that the shorter period of the ice coverage caused physical damages to facilities for aquaculture when drift ice surges into the lake through the channels. The lake normally becomes iced-over before the inflow so that it keeps the drift ice from surging into the lake. However, the shorter period of iced-over lake is likely to be occurred since 1968. Particularly, some years (in 1974, 1989, 1991, 1997, 2004 and 2007) experienced no iced-over lake. Once the drift ice surged into the lake, it brought physical damages to a large number of facilities for aquaculture. In fact, the fishers experienced a calamitous disaster related to this in 1974. Six to eight-meter block of drift ice beneath the water crashed their facilities, resulting in economic loss of around 2.3 billion Yen (Kato et al. 2004). The experience was a traumatic for the fishers, calling for urgent demands to the governments for building infrastructure development with an aim to keep drift ice from surging into the lake. Taking into account survey results conducted by academic researchers during 2 years from 1991 to 1992, the consensus building was made on constructing a sink-and-float style ice boom in Saroma Lake which is first installation in marine area (Kudo et al. 2000). The seasonal installation enabled fishers to manage scallop culture in a safety environmental condition. To implement and monitor the ice coverage and effects of the ice boom installation, resident researchers working in ACSL are in charge of disseminating the progress to the fishers, enabling to open fisheries management system against environmental and climate changes.

Second, the shortening of ice extent in Saroma Lake will have much influence on organic production due to the greater supply of light intensity, implying water degradation in the lake environment (Shiomoto 2008). In the complete iced-over lake, a decrease in organic production appears, resulting from a situation that ice and planktonic algae are exposed to the low light intensity. However, shorter periods and thinner of the ice coverage will promote higher light intensity into the lower part of the ice and the sea water under the ice. It is likely that the warming trend cause accelerated water contamination coupled with direct anthropogenic pressures from the upstream and scallop culture, requiring appropriate approaches to deal with water resource management in the region.

13.5 Community-Based and Scientific-Oriented Fisheries Management in Saroma Lake

Changes in the lake environment where greater warming is expected to occur may have a negative feedback on fishing production. Managing scallop culture should take account of profound interactions between fishing operation and their supporting ecosystem. Provided uncertainties and regional variability inherent in climate change, conventional community practices may not apply. In order to respond to changes in environmental and climate changes, the scientific community has the ability to play a leadership role in being an agent for solving emerging issues on the fisheries. On account of this, the fishers sought to develop ecosystem approach as the way forward to sustainable scallop production by applying community-based and scientific-oriented fisheries management with active involvements of both fishers and researchers.

Before ACSL was established, the lake water was divided and the three administrative authorizes take possessions of different water zones, causing difficulties for implementing appropriate management in the continuum ecosystem. However, the presence of ACSL led all the fishers belonging to three FCAs to build cooperative fishery governance system. Production of breeding and cultured scallops has been institutionalized under the banner of ACSL, enabling to adaptive fisheries management system in terms of utilization, culture and protection. It needs to be mentioned that not only did ACSL ensure a holistic approach in the management of the lake fisheries; it was able to fully utilize scientific expertise to provide scientific information for decision making. To promote stable fishing production in Saroma Lake, ACSL made endeavor to strengthen interactions between fishers and researchers; bridging dialogues and cooperation between the two; establishing an optimum management plans for responsible fisheries in the lake. In the latter, for example, a committee called Supporting Committee for Environmental Conservation in Saroma Lake was organized in 2001, in order to provide a clear understanding of where and how negative impacts are influenced in the ecosystem and biological production. Having recognized the significance of scientific oriented management, ACSL themselves started to employ full-time researchers (resident researchers) in 1987. The resident researchers, working on problems of interest to the fishers and searching workable the solutions, are expected to play an important role in linking community-based and scientific-oriented fisheries management. Their roles can be described in Fig. 13.4. They have major three tasks to be tackled.

First, they are expected to collect and analyze required data for fishing operation in order to disseminate their findings to the fishers. As for the scallop culture in Saroma Lake, major works done by the resident researchers include resource surveys of natural scallop spat, benthic scallops and diarrhetic shellfish poisoning, monitoring of growth and mortality rate in cultured scallops, the DNA analysis of scallops and water quality survey. These scientific findings are expected to serve as a fundamental basis for fishing operation to respond to environmental and climate changes. One of the best examples is collection of natural scallop spat. The fishers need to collect a large quantity of spat to operate the business in and around Saroma Lake.

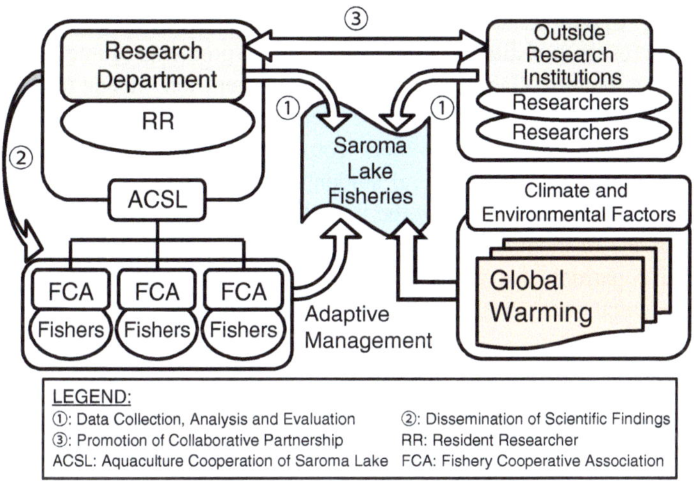

Fig. 13.4 Roles of resident researchers in Saroma Lake.

Taking into account appropriate timings on the basis of water temperature and size composition of the spat, each fisher stocks Saroma Lake with long-line bags with small-mesh size nets, which allow the scallop larvae to enter the mesh and adhere to the bags. The timings differ place to place and year to year, in response to external impacts such as water quality and global warming. In other words, no routine approach will be appropriate for the collection particularly in the recent situation. To address it, the resident researchers collect, analyze and disseminate to the members required data related to the spat during the period from April to August. The fishers, who received the information when appropriate, will be expected to start collection of natural scallop spat at just the right time and save boat navigation cost in an efficient manner.

Related to the above task, ACSL functions as an information center for the fishers to tackle with environmental and climate changes in the fisheries. Compared with conventional academic researchers who may give little or nothing back to those who donated their time and surrendered a measure of their privacy to the study, the resident researchers pay more attention to the way of how research is disseminated and used. To provide the fishers with the latest information including scallop resource survey, water quality monitoring and ice-coverage situation, news from ACSL are published and disseminated to the members while these can be contained on ACSL's homepage (http://www.saromako.sakura.ne.jp/centerhome.htm) on the internet. Around 57 issues of the news per year have been made on average during the last 5 years (2008–2012) (ACSL 2013). Besides, a series of work achievement reports have been published annually. The information will make great contributions to develop adaptive capacity to coevolve with resource and ecosystem dynamics in the short and long terms.

Third, the resident researchers play important roles in building strong partnership with outside research institutions. They working on problems of interest to the fishers can lead to better collaboration between the fishers and outside researchers as a catalyst. The information gathered by resident researchers can be used by not only the fishers but also relevant researchers from outside. Besides, visiting researchers including graduate and postdoctoral students can engage in certain research programs with assistance provided by the fishers through the resident researchers. The presence of the resident researchers is expected to create a platform for boosting collaborative partnerships among them and developing an adaptive strategy to respond to actual or expected impacts including climate change.

13.6 Conclusions

Putting them all together, this chapter highlighted adaptive fisheries management in Saroma Lake where temperature change has been rapid. The warming atmosphere is likely to impact negatively on the ice-covered ecosystem and fishing production. Duration of complete ice coverage tend to be shorter or not occurred, causing physical damages to facilities for aquaculture directly. Furthermore, the shortening is expected to stimulate drastic increases in phytoplankton growth called the spring bloom after disappearance of ice (Shiomoto et al. 2012; Fukuchi et al. 1989). Changes in primary produces commensurate with global warming can have pronounced effects on the structure and function of a given aquatic ecosystem (Shiomoto et al. 2012), possibly causing more pressures on the availability of fishery resources in the future.

Under the emerging risks coupled with climate change, the case study reveals that innovative efforts performed by the fishers and researchers have been made by applying community-based and scientific-oriented fisheries management in Saroma Lake. The community-based fisheries management are based on the establishment of ACSL. Three FCAs working in the lake applied cooperative approaches and the concept of common-pool resources (CPRs) was put into practice among them under the banner of ACSL. As a consequence, introduction and strict compliance of the Cultured Scallop Allowance Limit (CSAL) are one of significance achievements in implementing adaptive responses. Furthermore, the case study indicated that the resident researchers who were introduced by ACSL have contributed to yield a lot of benefits. To summarize, they have multiple roles in not only conducting relevant surveys by themselves, but also functioning as an information center and promoting better collaboration with outside researchers. Compared with common academic researchers who aim to report their findings in scholarly journals, the resident researchers require finding workable solutions to problems of interest to the fishers. Long-term commitments with the fishers can lead resident researchers to choose samples, methods, questions, observations and research designs that are required in resource management, enabling to respond to changes in the ecological-social-economic system with a process of trial and error. The useful information including monitoring of resource, water and ice coverage has been timely disseminated to the members in an appropriate manner. However, it needs to be mentioned that the resident researchers will face difficulties in producing information relevant to the range of practical research agenda which is complex and uncertain, in terms of limited knowledge and capacity. To overcome the challenges, the resident researchers strengthened its relationship with outside research institutions, in order to accumulate and fully utilize scientific knowledge for conserving the provisioning and regulatory services of the lake ecosystem. Permanent or long-term residency can allow the resident researchers to interpret concerns of the fishers and act in a leading role in the coordination of adaptive fisheries development in strong collaboration with outside stakeholders. Their roles will be more appreciate in the resource management debate particularly in the context of global warming.

Acknowledgements The author acknowledges the support of a JSPS grant and is extremely grateful to the staff of the Aquaculture Cooperation of Saroma Lake for their wholehearted support. The author is also thankful to local fishers and the staffs of fishery cooperative associations in Saroma Lake for taking time for interviews and field observation.

References

ACSL (1999a) Saroma-ko no kaze: rentai to kyousei, vol I. Aquaculture Cooperation of Saroma Lake, Tokoro (in Japanese)

ACSL (1999b) Saroma-ko no kaze: rentai to kyousei, vol II. Aquaculture Cooperation of Saroma Lake, Tokoro (in Japanese)

ACSL (2013) Homepage of Aquaculture Cooperation of Saroma Lake. http://www.saromako.org/. Accessed 15 July 2013 (in Japanese)

Allison EH, Perry AL, Badjeck MC, Adger WN, Brown K, Conway D, Halls AS, Pilling GM, Reynolds JD, Andrew NL, Dulvy NK (2009) Vulnerability of national economies to the impacts of climate change on fisheries. Fish Fisheries 10:173–196

Anthony A, Atwood J, August P, Byron C, Cobb S, Foster C, Fry C, Gold A, Hagos K, Heffner L, Kellogg OD, Lellis-Dibble K, Opaluch JJ, Oviatt C, Pfeiffer-Herbert A, Rohr N, Smith L, Smythe T, Swift J, Vinhateiro N (2009) Coastal lagoons and climate change: ecological and social ramifications in U.S. Atlantic and Gulf coast ecosystems. Ecol Soc 14(1). http://www.ecologyandsociety.org/vol14/iss1/art8. Accessed 15 July 2013

Armitage D, Berkes F, Doubleday N (eds) (2007) Adaptive co-management: collaboration, learning, and multi-level governance. UBC Press, Vancouver

Arrigo KR, Dijken GV, Pabi S (2008) Impact of a shrinking Arctic ice cover on marine primary production. Geophys Res Lett 35(L19603):1–6

Berkes F (2004) Rethinking community-based conservation. Conserv Biol 18(3):621–630

Berkes F (2005) Common theory for marine resource management in a complex world. Senri Ethnological Stud 67:13–31

Berkes F (2007) Adaptive co-management and complexity: exploring the many faces of co-management. In: Armitage D, Berkes F, Doubleday N (eds) Adaptive co-management: collaboration, learning, and multi-level governance. UBC Press, Vancouver

Brander K (2010) Impacts of climate change on fisheries. J Mar Syst 79:389–402

Carlsson L, Berkes F (2005) Co-management: concepts and methodological implications. J Environ Manage 75:65–76

FAO (2012) The state of world fisheries and aquaculture 2012. Food and Agriculture Organization, Rome

Fisheries Agency (2013) Gyogyou yoshoku gyou seisan toukei nenpou. http://www.maff.go.jp/j/tokei/kouhyou/kaimen_gyosei/index.html#y. Accessed 15 July 2013 (in Japanese)

Ford JD, Gough WA, Laidler GJ, MacDonald J, Irngaut C, Qrunnut K (2009) Sea ice, climate change and community vulnerability in northern Foxe Basin, Canada. Clim Res 38:137–154

Fukuchi M, Watanabe K, Tanimura A, Hoshiai T, Sasaki H, Satoh H, Yamaguchi Y (1989) A phytoplankton bloom under sea ice recorded with a moored system in lagoon Saroma ko, Hokkaido, Japan. Proc Natl Inst Polar Res Symp Polar Biol 2:9–15

ILEK (2013) ILEK homepage. http://www.en.ilekcrp.org/in. Accessed 15 July 2013

IPCC (2007) Climate change 2007: the physical science basis: summary for policymakers. http://www.ipcc.ch/pdf/assessment-report/ar4/wg1/ar4-wg1-spm.pdf. Accessed 15 July 2013

Iwasaki S, Shaw R (2010) Integrated lagoon fisheries management: resource dynamics and adaptation. Emerald Group Publishing, Bradford

Japan Metrological Agency (2013) Kako no kishou data kensaku. http://www.data.jma.go.jp/obd/stats/etrn/index.php. Accessed 15 July 2013 (in Japanese)

Kato S, Maekawa K, Sakaguchi K (2004) Lake Saroma, a rich cradle: its history and present conditions. In: Ramsar Center Japan (ed) Proceedings of the international workshop on wise use of Lagoon Wetlands, 23–25 July 2003, Kushiro (in Japanese)

Katsuki K, Seto K, Noguchi T, Sonoda T, Kim JY (2012) Effects of regional climate changes on the planktonic ecosystem and water environment in the frozen Notoro Lagoon, northern Japan. Mar Environ Res 81:83–89

Kudo M, Saeki H, Shimizu T, Takeda H, Ukegawa M, Kato S (2000) Drift-ice force affects the ice boom at the primary lake mouth of Saroma. Proc Hokkaido Tokai Univ Sci Eng 13:25–32 (in Japanese)

Ohshima KI, Mizuta G, Itoh M, Fukamachi Y, Watanabe T, Nabae Y, Suehiro K, Wakatsuchi M (2000) Winter oceanographic conditions in the southwestern part of the Okhotsk sea and their relation to sea ice. J Oceanogr 57:451–460

Perovich DK, Richter-Menge JA (2009) Annu Rev Mar Sci 1:417–441

Rijnsdorp AD, Peck MA, Engelhard GH, Möllmann C, Pinnegar JK (2009) Resolving the effect of climate change on fish populations. ICES J Mar Sci 66:1570–1583

Sato T, Imazu Y, Sakawa T, Kazama T, Wakabayashi T, Uhrenholdt T (2007) Modeling of integrated marine ecosystem including the generation-tracing type scallop growth model. Ecol Model 208:263–285

SCECSL (2005) Saromako no kankyo hozen wo support suru iinkai report. Supporting Committee for Environmental Conservation in Lake Saroma, Hokkaido (in Japanese)

Shiomoto A (2008) Long term trend of chemical oxygen demand in Saroma-ko Lagoon, Japan; possible effects of climatic warming. Nat Precedings. http://www.precedings.nature.com/documents/1857/version/1/files/npre20081857-1.pdf. Accessed 15 July 2013

Shiomoto A, Asakuma K, Hoon HD, Sakaguchi K, Maekawa K (2012) An early spring bloom of large diatoms in the ice-covered Saroma-ko Lagoon, Hokkaido, Japan. J Mar Biol Ass UK 92(1):29–37

Stroeve JC, Serreze MC, Holland MM, Kay JE, Malanik J, Barrett AP (2011) The Arctic's rapidly shrinking sea ice cover: a research synthesis. Clim Change 110:1005–1027

UNISDR (2007) Environment and vulnerability: emerging perspectives. The United Nations Environment Programme, Geneva

Vinnikov KY, Robock A, Stouffer RJ, Walsh JE, Parkinson CL, Cavalieri DJ, Mitchell JFB, Garrett D, Zakharov VF (1999) Global warming and northern hemisphere sea ice extent. Science 286:1934–1937

WWF (2005) Are we putting our fishy in hot water? http://www.assets.panda.org/downloads/fisherie_web_final.pdf. Accessed 15 July 2013

About the Editor

Rajib Shaw is a Professor in the Graduate School of Global Environmental Studies of Kyoto University, Japan. He worked closely with the local communities, NGOs, governments and international organization, including United Nations, especially in the Asian countries. He is currently the Chair of the United Nations Asia Regional Task Force for Urban Risk Reduction, and the President of Asian University Network of Environment and Disaster Management (AUEDM). His research interests are: community based disaster risk management, climate change adaptation, urban risk management, and disaster and environmental education. He has published several books in the field of disaster and environmental management. He is also the Chief Editor of Asian Journal of Environment and Disaster Management.

R. Shaw (ed.), *Community Practices for Disaster Risk Reduction in Japan*,
Disaster Risk Reduction, DOI 10.1007/978-4-431-54246-9, © Springer Japan 2014

MIX
Papier aus verantwortungsvollen Quellen
Paper from responsible sources
FSC® C105338
FSC
www.fsc.org

If you have any concerns about our products,
you can contact us on
ProductSafety@springernature.com

In case Publisher is established outside the EU,
the EU authorized representative is:
Springer Nature Customer Service Center GmbH
Europaplatz 3, 69115 Heidelberg, Germany

Printed by Libri Plureos GmbH
in Hamburg, Germany